Electronic and electrical servicing: Level 3

Consumer and Commercial Electronics core units

Electronic and electrical servicing: Level 3

Consumer and Commercial Electronics
core units

Ian Sinclair B.Sc.

Geoff Lewis B.A., M.Sc., M.R.T.S., M.I.E.E.

E E B

Newnes

OXFORD   AMSTERDAM   BOSTON   LONDON   NEW YORK   PARIS
SAN DIEGO SAN FRANCISCO SINGAPORE SYDNEY TOKYO

Newnes
An imprint of Elsevier Science
Linacre House, Jordan Hill, Oxford OX2 8DP
225 Wildwood Avenue, Woburn, MA 01801–2041

First published 2002

British Library Cataloguing in Publication Data
A catalogue record for this book is available from the British Library

ISBN 07506 55682

For information on all Newnes publications
visit our website at www.newnespress.com

Printed and bound in Malta

Contents

Preface

This new title has been written to meet the needs of the Core Units of the new Level 3 Progression Awards in Electrical and Electronics Servicing: Consumer and Commercial Electronics from the Electronics Examination Board and City & Guilds (6958). Based on the authors' previous texts published in the Servicing Electronic Systems series, in particular Volume 2, Part 1, this title is also ideal as a text for students following City & Guilds 2240 Part 2. In addition, this text provides the underpinning knowledge needed for Level 3 NVQs.

Acknowledgements

The authors would like to thank the Electronics Examination Board (EEB) for their endorsement of this book, and the companion title for Level 2 as suitable texts for the new Progression Awards. They also gratefully acknowledge the help provided by the EEB, City & Guilds, and EMTA in the development of this book.

We would also like formally to recognize the contributions of the many lecturers and course tutors who have made many useful and constructive suggestions over the past twenty years, during the development of this series and its predecessors, Servicing Electronic Systems and Electronics for the Service Engineer.

Ian Sinclair

Geoff Lewis

April 2002

Unit 1

Outcomes

1. Demonstrate an understanding of reactance, resonance and transformers together with the practical applications of such circuits and components.
2. Demonstrate an understanding of semiconductor devices, displays and transducers, and the practical applications of these components.

Note

This unit provides the underpinning knowledge associated with Optional units, 5, 8, and 10 of the Electrical and Electronics Servicing at Level 3.

1 Sine wave driven circuits

Reactance

Capacitors and inductors in d.c. circuits cause transient current effects only when the applied voltage changes. In an a.c. circuit analysis, this is considered to be a continuous process. Capacitors in such a circuit are therefore continually charging and discharging, and inductors are continually generating a back-e.m.f. Circuits containing resistors, capacitors and inductors are described as *complex circuits* and an alternating voltage will exist across each component proportional to the magnitude of current flowing through it so that a form of Ohm's law still applies.

In the explanation that follows, the symbols V' and I' are used to mean a.c. values (peak or r.m.s.) of alternating signals. The symbols V and I will have their usual meaning of d.c. values.

In a resistor, $V' = RI'$; and the value of resistance found from the variant form of this equation, $R = V'/I'$, is the same as the d.c. value, V/I. In a capacitor or an inductor, the ratio V'/I' is called *reactance*, symbolized as X. Because this is a ratio of volts to amperes, the same units (ohms) are used to express its *resistance*, but this resistance is a quantity of a very different kind.

A capacitor may, for example, have a reactance of only 1K, but a d.c. resistance that is unmeasurably high. An inductor may have a d.c. resistance of 10 ohms, but a reactance of 5K or more. The reactance of a capacitor or an inductor is not a constant quantity but depends upon the frequency of the applied signal. A capacitor, for example, has a very high reactance to low frequency signals and a very low reactance to high-frequency signals.

Table 1.1 Capacitive reactance at audio frequencies

Value	20 Hz	50 Hz	400 Hz	1 kHz	5 kHz	10 kHz	20 kHz
470 pF	∞	∞	847K	339K	68K	34K	17K
2n2	∞	∞	181K	72K	15K	7K	3K6
10n	796K	318K	40K	16K	3K	1K6	798R
47n	16K	68K	8K5	3K4	678R	339R	169R
220n	36K	14K	1K8	724R	145R	72R	36R
1 µF	8K	3K	400R	160R	32R	16R	8R
10 µF	800R	318R	40R	16R	0	0	0
100 µF	80R	32R	4R	Low	0	0	0

Note. All figures have been rounded. Very high or low values have been shown as open (∞) or short (0) circuits.

The reactance of a capacitor V/I can be calculated from the equation:

$$X_c = 1/(2\pi \times f \times C) \text{ ohms}.$$

where f is the frequency of the signal in Hz and C is the capacitance in farads. Tables 1.1 and 1.2 show the values of capacitive reactance for a range of different capacitors calculated for a range of frequencies. These tables are intended as a guide so that you can quickly estimate a reactance value without the need to make the calculations.

Table 1.2   Capacitive reactance at radio frequencies

Value	100 kHz	470 kHz	1 MHz	5 MHz	10 MHz	20 MHz	30 MHz
10 pF	160K	34K	16K	3K	1K6	800R	531R
22 pF	72K	15K	7K	1K5	724R	362R	141R
100 pF	16K	3K4	1K6	320R	160R	80R	53R
470 pF	3K4	721R	339R	68R	34R	17R	Low
1 nF	1K6	339R	160R	32R	16R	Low	Low
4n7	339R	72R	34R	Low	0	0	0
22 nF	72R	15R	7R	0	0	0	0
47 nF	34R	7R	0	0	0	0	0

Note. All figures have been rounded. Very high or low values have been shown as open (∞) or short (0) circuits.

Question 1.1

Calculate the reactance of a 22 nF capacitor at a frequency of 2 kHz.

Activity 1.1

Using the data of Table 1.2, calculate the reactances for each frequency for a capacitor of 100 pF and then draw and comment on the shape of the graph of reactance plotted against frequency. Answers in Appendix 4.

The reactance of an inductor varies in the opposite way, being low for low frequency signals and high for high-frequency signals. Its value can be calculated from the equation:

$$X_L = (2\pi \times f \times L) \text{ ohms}.$$

where f is the frequency in Hz and L is the inductance in henries. Tables 1.3 and 1.4 show the values of inductive reactance which are found at various frequencies.

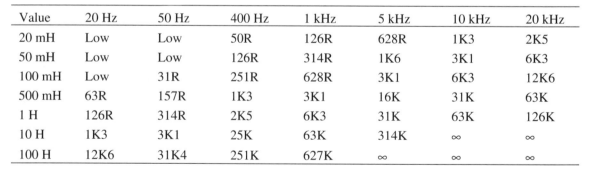

Activity 1.2

Note: inductors are now less common in circuits other than power supply and radio (transmission and reception) applications. The increasing use of ICs and digital circuitry has made the use of inductors unnecessary in a very wide range of modern applications.

Connect the circuit shown here, left. If meters of different ranges have to be used, changes in the values of capacitor and inductor will also be necessary. The signal generator must be capable of supplying enough current to deflect the current meter which is being used.

Connect a 5 µF capacitor between the terminals, and set the signal generator to a frequency of 100 Hz. Adjust the output so that readings of a.c. voltage and current can be made. Find the value of V/I at 100 Hz.

Repeat the readings at 500 Hz and at 1000 Hz. Tabulate values of $Xc = V/I$ and of frequency f.

Now remove the capacitor and substitute a 0.5 H inductor. Find the reactance at 100 Hz and 1000 Hz as before, and tabulate values of $X_L = V/I$ and of frequency f.

Next, either remove the core from the inductor or increase the size of the gap in the core (if this is possible), and repeat the measurements. How has the reactance value been affected by the change?

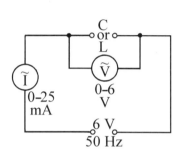

Table 1.3   Inductive reactance tables for audio frequencies

Value	20 Hz	50 Hz	400 Hz	1 kHz	5 kHz	10 kHz	20 kHz
20 mH	Low	Low	50R	126R	628R	1K3	2K5
50 mH	Low	Low	126R	314R	1K6	3K1	6K3
100 mH	Low	31R	251R	628R	3K1	6K3	12K6
500 mH	63R	157R	1K3	3K1	16K	31K	63K
1 H	126R	314R	2K5	6K3	31K	63K	126K
10 H	1K3	3K1	25K	63K	314K	∞	∞
100 H	12K6	31K4	251K	627K	∞	∞	∞

Note. All values have been rounded. At extremely high frequencies an inductor behaves as an open circuit (symbolised as ∞), and at extremely low frequencies, an inductor behaves as a low resistance.

Table 1.4 Inductive reactance tables for radio frequencies

Value	100 kHz	470 kHz	1 MHz	5 MHz	10 MHz	20 MHz	30 MHz
10 μH	Low	30R	63R	314R	628R	1K3	1K9
50 μH	31R	148R	314R	1K6	3K1	6K3	9K4
200 μH	126R	590R	1K3	6K3	12K6	25K	37K7
1 mH	630R	2K95	6K5	31K5	63K	125K	188K5
2 mH	1K3	6K	12K6	63K	126K	251K	High
5 mH	3K1	14K8	31K4	157K	315K	High	V-high
10 mH	6K3	29K	63K	315K	High	V-high	V-high

Note. All values have been rounded. At extremely high frequencies an inductor behaves as an open circuit (symbolised as ∞), and at extremely low frequencies, an inductor behaves as low resistance.

Activity 1.3

Calculate the values for the inductive reactance for the 1 mH component in Table 1.4 and then draw and comment of the graph of reactance *vs* frequency. Answers at the end of Appendix 4.

Phase angle

current I
V resistor
V capacitor
V inductor

There is another important difference between a resistance and a capacitive or inductive reactance and this can be demonstrated as follows.

With the aid of a double-beam oscilloscope, the a.c. waveform of the current flowing through a resistor and the voltage developed across it can be displayed together (see illustration, left). This shows, that the two waves coincide, with the peak current coinciding with the peak voltage, etc. If this experiment is repeated with a capacitor or an inductor in place of the resistor, you will see from the figure that the waves of current and voltage do *not* coincide, but are a quarter cycle (90°) out of step.

Comparing the positions of the peaks of voltage and of current, you can see that:

- With a capacitor, the current wave leads (or precedes) the voltage wave by a quarter-cycle.
- With an inductor, the voltage wave leads the current wave also by a quarter-cycle.

An alternative way of expressing this is, for a capacitor, the voltage wave *lags* (or arrives after) the current wave by a quarter-cycle, and for an inductor, the current wave lags the voltage wave by a quarter-cycle.

The amount by which the waves are out of step is usually defined by the phase angle. The reason is that a coil of wire rotating in the field of a magnet generates a sine wave, with one cycle of wave being generated for every turn (360°) of rotation. One half-cycle thus corresponds to 180°, and one quarter-cycle to 90°. Thus the current and voltage waves are 90° out of phase in a reactive component such as a capacitor or an inductor.

- A useful way to remember the phase relationship between the current and voltage is the word C-I-V-I-L, meaning C — I leads V; V leads I — L. The letters C and L are used to denote capacitance and inductance respectively.

Question 1.2

The phase shift between two waves is 45°. If the frequency of each wave is 5 kHz, what will be the time difference between corresponding peaks as observed on the oscilloscope?

Phasor diagrams

It is necessary only to take a few measurements on circuits containing reactive components to see that the normal circuit laws used for d.c. circuits cannot be applied directly to a.c. circuits.

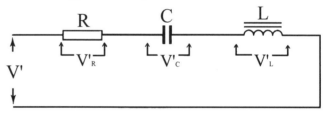

Figure 1.1   Circuit containing reactive components

Consider, for example, a series circuit containing a 10 μF capacitor C, a 2 H inductor L and a 470 ohm resistor R, as in Figure 1.1. With 10 V a.c. voltage V' at 50 Hz applied to the circuit, the a.c. voltages across each component can be measured and added together ($V'_C + V'_L + V'_R$). You will find that these measured voltages do not add up to the voltage V' across the whole circuit.

Activity 1.4

Connect the circuit shown, left, with the values given above. Use either a high-resistance a.c. voltmeter or an oscilloscope to measure the voltage V'_R across the resistor and the voltage V'_C across the capacitor. Now measure the total voltage V', and compare it with $V'_R + V'_C$.

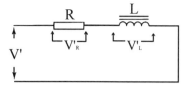

Repeat the procedure, substituting the inductor L for the capacitor C and finding V', V'_R and V'_L. Again compare V' with $V'_R + V'_L$.

The reason why the component voltages in a complex circuit do not add up to the circuit voltage when a.c. flows through it is due to the phase angle between voltage and current in the reactive component(s). At the peak of the current wave, for example, the voltage wave across the resistor will also be at its peak, but the voltage wave across any reactive component will be at its zero value. Measurements of voltage cannot however, indicate phase angle. They can only give the r.m.s. or peak values for each component, and the fact that these values do not occur at the same time cannot be allowed for by meter measurement. The result is that straight addition of the measured value will inevitably give a wrong result for total voltage, because of the time difference.

Phasor diagrams (often also called *vector* diagrams) are one method of performing the addition so that phase angle is allowed for. In a phasor diagram, the voltage across a resistor in an a.c. series circuit is represented by the length of a horizontal line drawn to scale. Voltages across reactive components are represented by the lengths of vertical lines, also drawn to the same scale. If all the lines are drawn from a single point, as in Figure 1.2(a), the resulting diagram is a phasor diagram that represents both the phase and the magnitude of the voltage wave across each component.

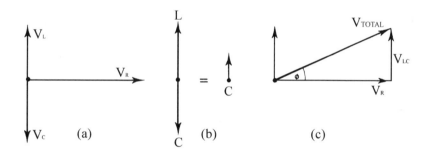

Figure 1.2 Phasor diagrams for complex series circuits

To represent the opposite effects that capacitors and inductors have on the phase, the vertical line representing voltage across an inductor is drawn vertically upwards, and the line representing voltage on a capacitor is drawn vertically downwards. (By convention, inductive reactance and hence the voltage across it are considered to be positive. Capacitive reactance and its voltage are thus considered to be negative.)

The phasor diagram can now be used to find the total voltage across the whole circuit. First, the difference between total upward (inductive) and total downward (capacitive) voltage is found, and a line is drawn to represent the size and direction of this difference.

For example, if the inductive voltage is 10 V and the capacitive voltage 7 V, the difference is 3 V drawn to scale in the direction of inductive reactance. If the inductive voltage were 10 V and the capacitive voltage 12 V, the difference would be 2 V drawn to scale downwards in the capacitive direction.

The net reactive voltage so drawn is then combined with the voltage across the resistor in the following way. Starting from the point marking the end of the line representing the voltage across the resistor, draw a vertical line, as in Figure 1.2(b), to represent the net reactive voltage in the correct direction, up or down. Then connect the end of this vertical line to the starting point, Figure 1.2(c).

The length of this sloping line will give the voltage across the whole circuit, and its angle to the horizontal will give the phase angle between voltage and current in the whole circuit.

Impedance

A complex circuit which contains both resistance and reactance possesses another characteristic which is of great importance. This characteristic is known as *impedance*, symbolized by Z. Impedance is measured in ohms, and is equal to the quotient of the values, V'/I', for the whole circuit. Its value varies as the frequency of the signal varies.

When impedance is present, the phase angle between current and voltage is less than 90°. This phase angle can be found most easily by using the phasor diagram in a slightly different way, to form what is known as the impedance triangle.

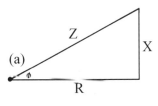

(a)

In a phasor diagram constructed with this object, left, separate lines are drawn to represent the resistance R, the reactance X and the impedance Z. In a series circuit, the length of the horizontal line represents the total value of resistance in the circuit, and the vertical line its net value of reactance upwards as before, for predominantly inductive reactance, (a) and downwards for predominantly capacitive reactance, (b).

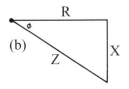

(b)

With the values of R and X known and the angle between them a right angle the Z line can be drawn in, representing the impedance value of the whole circuit. The angle of this line to the horizontal is the phase angle between current and voltage in the circuit.

Another way of working out the relationships between R, X and Z in a complex circuit is to express them by two algebraic formulae:

$$Z = \sqrt{((X_L - X_C)^2 + R^2)} \qquad \text{and} \qquad \tan \phi = (X_L - X_C)/R$$

where Z = total impedance, X_L = inductive reactance, X_C = capacitive reactance, and R = the resistance of the circuit as a whole. A pocket calculator covering a reasonably full range of mathematical functions can now be used to work out the values of circuit impedance and phase angle respectively.

Question 1.3

Find by any method the impedance of a circuit whose reactance is 5K with resistance 12K.

Filters

Filters are circuits that are designed to separate out a range of frequencies of interest that are present in any waveband. Because these are frequency dependent, the circuits must contain at least one reactive component.

In the low pass filter (LPF) shown in Figure 1.3(a), the inductor has a low series reactance at low frequencies so that these pass easily to the output and are developed as the output signal V_{out}.

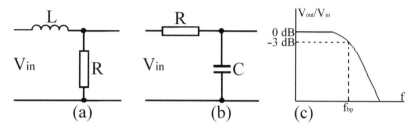

Figure 1.3   Low-pass filters (a) LR, (b) CR, (c) typical response

In Figure 1.3(b), the capacitor has a very high reactance at low frequencies and acts as the load to develop the output signal V_{out}. At high frequencies the low reactance of C effectively short circuits the output signal.

Figure 1.3(c) shows the amplitude of the output signal plotted against frequency for both the LR and CR circuits and this also represents the variation of circuit impedance with frequency. The -3 dB or half power point of the frequency response represents the so called *break point* where the circuit reactance is equal to its resistance value. This point is defined by the formula, either

$$f = 1/(2\pi CR) \text{ Hz or } f = 1/(2\pi(L/R)) \text{ Hz}$$

and is therefore dependent upon the circuit *time constant*.

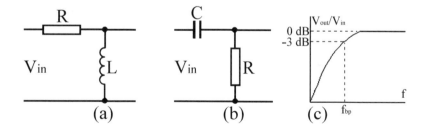

Figure 1.4   High-pass filters (a) LR, (b) CR, (c) typical response

For the high pass filters (HPF) shown in Figure 1.4(a and b), the resistors and reactors are interchanged to produce the opposite effect. Figure 1.4(c) again shows the attenuation effects and the variation of circuit impedance, with the break point calculated as before.

For both these *1st order* LPF and HPF circuits with a single pair of components shown above, the straight part of the attenuation slope falls away at 6 dB per octave (a doubling or halving of frequency).

Activity 1.5

For a LPF and HPF as shown in Figures 1.3 and 1.4, set up a circuit so that the signal current through the filter can be plotted against frequency. How do these results compare with the shape of the attenuation characteristics.?

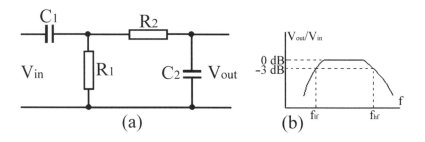

Figure 1.5   (a) Band-pass filter circuit, (b) typical response

A circuit that selects a band of frequencies is described as a *band pass filter* (BPF) and one such circuit can be constructed as shown in Figure 1.5(a). This is effectively a cascade of a LPF and a HPF and its attenuation characteristic is shown in Figure 1.5(b). The attenuation slope at both ends is still 6 dB per octave. The two –3 dB break points are calculated from the equations,

$$f_{lf} = 1/(2\pi C_1 R_1) \text{ Hz and } f_{hf} = 1/(2\pi C_2 R_2) \text{ Hz.}$$

The bandwidth of the circuit is simply the difference between these two frequencies.

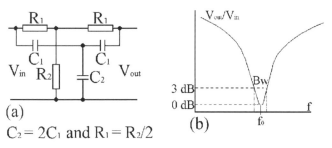

$C_2 = 2C_1$ and $R_1 = R_2/2$

Figure 1.6   Band-stop filter (twin-T) circuit (a) and typical response (b)

A band stop filter (BSF) which has the opposite characteristic of the band pass filter cannot be conveniently constructed using the cascading principle shown above. For these circuits it is more usual to employ 2nd order filters as shown in Figure 1.6(a). This particular circuit is referred to as a *twin* or *parallel T* device where the ratio of the component values are chosen as:

$$C_2 = 2C_1 \text{ and } R_1 = R_2/2 ,$$

so that the maximum attenuation occurs at f = $1/(2\pi R_1 C_1)$ Hz.

Resonance

Part of the process of calculating the value of an impedance involves, finding the difference between the values of capacitive reactance, X_C, and inductive reactance, X_L. At low frequencies, X_C is large and X_L small, a situation that becomes reversed at high frequencies. There must therefore be some frequency at which $X_C = X_L$.

This frequency is called the resonant frequency, or the frequency of resonance, of the LCR circuit in question. Its symbol is f_r.

Notes.

(a) The resonant frequency of a circuit containing inductance and capacitance is that at which the effects of the two reactances are equal.

(b) Since the reactance values are drawn in opposite directions, they cancel out on the phasor diagram.

(c) The impedance of a series LCR circuit at resonance is equal to its resistance value only.

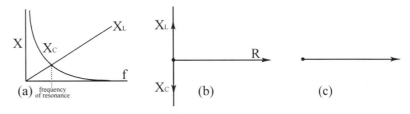

Figure 1.7   Resonance in a series LCR circuit

A phasor diagram drawn for a series LCR circuit at its resonant frequency will clearly have a zero vertical component of reactance (Figure 1.7). The impedance of the circuit will therefore be simply equal to its resistance. The same conclusion can be reached by working out the impedance formula:

$$Z = \sqrt{(X_L - X_C) + R^2}$$

At its resonant frequency, therefore, an LCR circuit behaves as if it contained only resistance, and has zero phase angle between current and voltage.

Activity 1.6

Connect the circuit shown, left, with component values as follows: R = 1K, C = 0.1 µF, and L = 80 mH. The resonant frequency of the circuit is about 1.8 kHz.

Set the signal generator to 100 Hz, and connect the oscilloscope so as to measure the voltage across the resistor R. This voltage will be proportional to the amount of current flowing through the circuit, because $V = R \times I$.

Now increase the frequency, watching the oscilloscope. The resonant frequency is the frequency at which current flow (and therefore the voltage across R) is a maximum. Note this frequency, and the value of the amplitude of the voltage across R at the resonant frequency.

Measure the voltages across L and across C by connecting the oscilloscope across each in turn. Note the value of these voltages.

Finally, use the oscilloscope to measure the voltage across the whole circuit.

Construct a phasor diagram for the voltages across R, C and L and confirm that this produces an answer for the total voltage. (Remember that the oscilloscope itself will disturb the circuit to some extent, and that the resistance of the inductor has not been taken into account in your calculation.)

Series resonance

In a *series-resonant circuit,* the current flow at resonance in the circuit will be large if the voltage across the whole circuit remains constant. Therefore a large voltage will exist across each of the reactive components. Generally, the voltage across both the capacitor and the inductor will be greater than the voltage across the whole circuit at the resonant frequency.

The ratio; V'_x/V'_z, where $V'x$ is the voltage across a reactor and $V'z$ is the voltage across the whole circuit, is called the *circuit magnification factor* Q, which can be very large at the frequency of resonance.

The frequency of resonance for a series circuit can be calculated by using the formula:

$$f_r = \frac{1}{2\pi\sqrt{LC}}$$

where L is the inductance in henries, C the capacitance in farads, and f the frequency in hertz.

Example. What is the resonant frequency of a circuit containing a 200 mH inductor and a 0.05 µF capacitor?

Solution. Substitute the data in the equation taking care to reduce both L and C to henries and farads respectively. Using $L = 200 \times 10^{-3} = 0.2$ H, and $C = 0.05 \times 10^{-6} = 5 \times 10^{-8}$ F. Take 2π as being approximately 6.3. Then:

$$f_r = 1/(6.3\sqrt{(0.2 \times 5 \times 10^{-8})}) = 1587 \text{ Hz}.$$

Question 1.4

Find (in MHz, to two decimal places) the resonant frequency of the series combination of a 220 pF capacitor and a 15 µH inductor.

Parallel resonance

A circuit consisting of inductance, capacitance and resistance in parallel will resonate at a frequency given approximately by the same equation that was used for series resonance. However, in a practical circuit, the resistive component is more likely to be the inherent resistance of the inductor. In which case the circuit now consists of C in parallel with L plus a series resistance R that represents this component's losses. The frequency of resonance for this circuit is given by:

$$f_r = \frac{1}{2\pi}\sqrt{\frac{1}{LC} - \frac{R^2}{L^2}}$$

If R is small as is commonly the case, the R^2/L^2 component can be neglected and the resonance frequency formula reverts to that for the series case. The lossy term R^2/L^2 is related to the Q factor (or quality factor, $Q = 2\pi f L/R$) of the inductor so that if Q is higher than about 50 the simplified formula is quite accurate enough for most applications.

At the frequency of resonance, a parallel resonant circuit behaves like a high value of resistance, $L/CR\ \Omega$, which is called the dynamic resistance or impedance. Again, at resonance, the phase angle between voltage and current is zero.

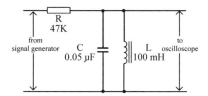

Activity 1.7

Connect the parallel resonant circuit shown, left, to the signal generator. Find the frequency of resonance, which for the component values shown will be about 2250 Hz, and note that at this resonant frequency, the voltage across the resonant circuit is a maximum.

Now connect another 0.05 μF capacitor in parallel with C, and note the new frequency of resonance.

Remove the additional capacitor, and plot a graph of the voltage across the resonant circuit against frequency, for a range of frequencies centred about the resonant frequency.

Observe the shape of the resulting curve, which is called the *resonance or response curve*.

Now add a 10K resistor in parallel with the resonant circuit, and plot another resonance curve, using the same frequency values. What change is there in the shape of the curve?

Repeat the experiment using a 1K resistor in place of the 10K one, and plot all three graphs on the same scale.

The result of these experiments will show that the addition of either capacitance or inductance to a parallel resonant circuit causes the frequency of resonance to become lower. The addition of resistance in parallel has little effect on the frequency of resonance, but a considerable effect on the shape of the resonance curve. The effect of adding a small value of resistance is to lower the peak of the resonance curve as might be expected

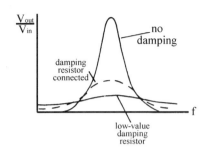

because the sum of two resistors in parallel is a net resistance smaller than either. In addition, however, the width of the curve is increased.

A resistor used in this way is called a *damping resistor*. Its effect is to make the resonant circuit respond to a wider range of frequencies, though at a lower amplitude. A damping resistor therefore increases the bandwidth of a resonant circuit, making the circuit less selective of frequency. When a parallel resonant circuit is used as the load of an amplifier, the tuned frequency is the resonant frequency of the parallel circuit, and the amount of damping resistance employed will determine the bandwidth of the amplifier.

Using LCR circuits allows us to construct band-pass and band-stop filters which are more efficient than those described above. The simple LC parallel or series circuits can have resistors added to dampen the resonance so that the resonant effect is reduced but spread over a wider range of frequencies, providing a simple band-pass or band-stop action according to where the resonant circuit is placed.

For many purposes, however, these circuits do not provide a sharp enough distinction between the pass and the stop bands and much more elaborate filter circuits have to be devised. Band-pass and band-stop characteristics are often plotted on a linear scale of frequency so that the bandwidth is easier to read from the graph, but the attenuation is always plotted in terms of decibels.

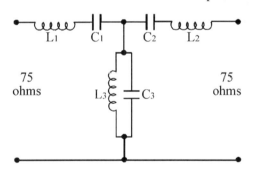

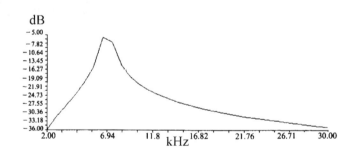

Figure 1.8   The circuit (a) and the computed response curve (b) for a band-pass filter

Calculations on such filters are very difficult, but computer programs can be used to print out a graph of response for any combination of components. Figure 1.8 shows a band-pass filter, using a parallel circuit to determine the band-pass mid-frequency. The graph shows the response of this circuit for the components as follows:

Source and load impedance		75Ω	
L_1, L_2	24 mH	L3	5 mH
C_1, C_2	33 nF	C_3	120 nF

Note that the graph produced by the computer program shows that there is some attenuation even at the pass-band. This is because the computer takes account of the input and output resistances which can not always be ignored.

Answers to questions

1.1	3K6 (3617 ohms).
1.2	25 μs.
1.3	13K.
1.4	2.79 MHz.

2 Transformers and power transfer

Transformers

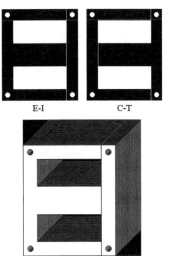

E-I　　　C-T

A transformer consists of two or more inductors so wound that their magnetic fields interact. (Note – the auto-transformer which is described below is a variation on this principle.) The mutual interaction is achieved by winding the complete set of coils on a common magnetic core that consists of either a stack of E-I or C-T soft iron laminations or a toroidal shaped ferrite core. The windings are referred to as the primary (input) and secondaries (outputs) respectively. For audio applications or low frequency power supply units, the cores may be made either of ferrite, silicon iron alloy or *mumetal*. This latter is an alloy that consists chiefly of nickel, iron, copper, manganese and chromium. For higher radio frequency applications, the coils may be either air or ferrite cored. In the latter, case the position of the ferrite slug may be adjusted in order to tune the circuits to resonance. Most of the modern power supply units operate on the switched mode principle and these use transformers that are wound on toroidally-shaped ferrite cores.

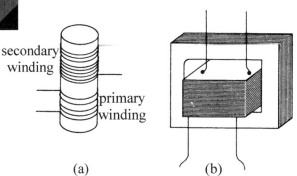

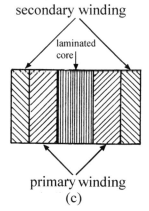

(a)　　　　　(b)　　　　　(c)

Figure 2.1 Transformer types and symbols

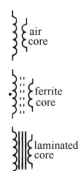

Some typical construction methods for transformers are illustrated in Figure 2.1. Type (a) would be used at radio frequencies; type (b) at lower frequencies with a cross-section through the construction shown at (c). The symbols, left, are used in circuit diagrams to indicate various types of transformer core.

The inputs may be provided either by a mains driven power system or other signals, that may be either pure a.c. or varying level unidirectional current. These are applied to the primary winding and induce output signals, which are *always* a.c., across the terminals of the secondary winding(s). Since the mutual magnetic effect depends entirely on a changing input voltage level, a steady d.c. input current in the primary winding would induce zero output signal in the secondary windings.

When the current is taken from the secondary winding(s) by connecting a load, an increased primary current must flow to provide the power which is being dissipated. If no current is taken from the secondary winding, the residual current flow in the primary winding, described as the *magnetizing current*, will be very small.

The input and output signal voltages may be in-phase or anti-phase depending upon the polarity of the secondary connections and the relative directions of the windings. The secondaries may consist of a single winding, multiple separate windings, or a single winding with multiple voltage taps each designed to provide a particular level of output voltage. Some secondary windings may be centre tapped to provide balanced output voltages.

Ideal transformer

The ideal transformer would be one which has no loss of power when in use, so that no primary current at all would flow until a secondary current was being drawn. Large transformers in fact approach quite close to this ideal which is used for the basic transformer calculations. In an ideal transformer where V'_s = secondary a.c. voltage, V'_p = primary a.c. voltage; n_s = number of turns of wire in the secondary winding, and n_p = number of turns of wire in primary winding:

$$\frac{V'_s}{V'_p} = \frac{n_s}{n_p}.$$

Example. A transformer has 7500 turns in its primary winding and is connected to a 250 V 50 Hz supply. What a.c. voltage will be developed across its secondary winding if the latter has 500 turns?

Solution. Substituting the data in the formula;

$$V'_s/V'_p = n_s/n_p,$$

Then $V'_s/250 = 500/7500 = 1/15$, so that $V'_s = 250/15 = 16.67$ or nearly 17 volts.

In practice, because no transformer is perfect, the output voltage would probably be somewhat less than 16 V.

Question 2.1

A step-down transformer uses a 4:1 ratio. If the primary voltage is 240 V a.c., what (assuming no losses) is the secondary output?

In the ideal transformer, the power input to the primary winding must be equal to the power taken from the secondary winding, so that

$$V'_p \times I'_p = V'_s \times I'_s \text{ or, rearranging, } \frac{V'_s}{V'_p} = \frac{I'_p}{I'_s}.$$

since $\dfrac{V'_s}{V'_p} = \dfrac{n_s}{n_p}$, it then follows that $\dfrac{I_p}{I_s} = \dfrac{n_s}{n_p}$ or $I_p \times n_p = I_s \times n_s$

This last equation is often a convenient form which relates the signal currents in the perfect transformer to the number of turns in each of the two windings.

Transformers are used in electrical circuits for the following purposes:

1. Voltage transformation – converting large signal voltages into low voltages, or vice versa, with practically no loss of power.
2. Current transformation – converting low-current signals into high-current signals, or vice versa, with practically no loss of power.
3. Impedance transformation – enabling signals from a high-impedance source to be coupled to a low impedance, or vice versa, with practically no loss of power through mismatch.
4. Electrical isolation – for service purposes such a transformer positioned between the mains supply and any secondary loads will avoid electrical shocks to the operator.

Note that the transformer is a passive device without power gain. If a transformer has a voltage step-up of ten times, it will also have a current step-down of ten times (assuming no losses en route).

Example. The secondary winding of a transformer supplies 500 V at 1 A. What current is taken by the 250 V primary?

Solution. Since $V'_s.I'_s = V'_p.I'_p$, then $250 \times I'_p = 500 \times 1$, so that $I'_p = 2$ A.

Question 2.2

A transformer operates from 240 V mains and delivers an output of 60 W at 6 V. Assuming no losses, what amount of primary current flows?

Transformers may also be used as matching devices so that the maximum power can be transferred from one circuit to another. The ideal method of delivering power to a load would be to use amplifiers which have a very low internal resistance, so that most of the power (I^2R) was dissipated in the load. Many audio amplifiers make use of such transistors to drive 8 ohm loudspeaker loads. For some purposes, however, transistors that have higher resistance must be used or loads that have very low resistance must be driven, and a transformer must be used to match the differing impedances.

In public address systems, for example, where loudspeakers are placed at considerable distances from the amplifier, it is normal to use 100 volt line signals at low currents so as to avoid I^2R losses in the network. In such cases the 8 ohm loudspeakers must be coupled to the lines through transformers.

From the voltage, current and turns ratios described above, we can deduce that the input and output load resistance values have the following relationship:

$R_s/R_p = (n_s/n_p)^2$ which can be re-arranged as

$R_p = (n_p/n_s)^2 \times R_s$.

so that the equivalent input resistance to signals entering a perfect transformer is $R_L(n_p/n_s)^2$, where R_L is the load resistance connected to the secondary winding. The equivalent circuit for a perfect transformer is therefore that shown in Figure 2.2.

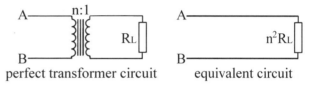

perfect transformer circuit   equivalent circuit

Figure 2.2 Equivalent circuit of a perfect transformer

For the maximum transfer of power, the formula for the turns ratio $n = n_p/n_s$ can be expressed in the following way,

$$n = \sqrt{\frac{\text{amplifier output impedance}}{\text{impedance of load}}}$$

Example: A power amplifier stage operates with a 64 ohm output impedance. What transformer ratio is needed for maximum power transfer to an 8 ohm load?

Solution: A 3:1 step-down transformer could therefore be used off-the-shelf, or a transformer specially wound for a 2.8:1 ($\sqrt{8}$) ratio.

Question 2.3

A transformer is to be used to match a source resistance of 100 ohms to a loudspeaker of 4 ohms. What ratio is needed?

Activity 2.1

Set up the circuit shown in Figure 2.3 using a multi-ratio output transformer. Tr1 can be either a 2N3053 or a BFY50, Tr2 a 2N3055 and D, a 1N4001. With a 40 V supply, adjust VR1 so that the standing d.c. current flow through Tr2 is 50 mA. Connect the 8 ohm resistor between two of the taps of the secondary of the transformer and connect the signal generator to the input. Connect an oscilloscope across the 8 ohm resistor.

Applying a 400 Hz signal, adjust input signal voltage so that a 2 V peak-to-peak signal is observed across the 8 ohm resistor. Now, without altering the signal generator settings, switch off the amplifier, change the transformer connections so as to alter the ratio, and switch on again. Measure the output voltage.

Repeat the procedure so that the output voltage is measured for every possible pair of secondary tapping points. What transformer ratio gives the maximum signal output? Is the setting critical?

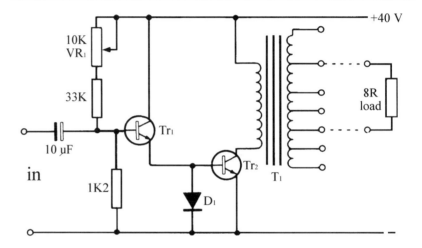

Figure 2.3   Circuit for Activity 2.1

Transformer power ratings

Due to the nature of the self inductance and capacitance of the transformer and the effects of the load, the a.c. voltage and current in the secondary circuit are rarely in phase. The power loading measured in watts can therefore be misleading and a more meaningful assessment uses a *volt amps* (VA) relationship. Furthermore, the rising volts drop that occurs as the load current increases, is described by the *regulation factor*. Typically in small to medium sized equipment power units, this accounts for losses of about 10%.

Consider the calculations associated with the following transformer designed to provide a secondary supply of 12 volts at 4 amps or 48 VA. This would produce an output of 12 V when driving a 48 watt load. As the load is reduced, the voltage will rise due to the regulation factor by about 10% to 13.2 V. (Note that this is not the same as the d.c. output after rectification.)

Using a typical value of 4.8 turns per volt plus an extra 1% for each 10 VA of loading produces $12 \times 4.8 + (1\% \text{ of } 4.8)$ or approximately 60 secondary turns. Looking up wire tables would show that this loading could be safely supported by 1.25 mm diameter wire.

For the 250 V primary winding at 60 turns/12 V (or 5 turns/volt), this will require $250 \times 5 = 1250$ turns.

Maximum transfer of power

If a generator and load are both resistive (V and I in phase), then the maximum transfer of power occurs when the internal generator resistance (R_G) and load resistance (R_L) are equal (i.e. when $R_G = R_L$). Matching of these values can be achieved using a transformer.

When the generator or load represents an impedance (V and I not in phase), maximum transfer of power occurs when the magnitudes of the impedances are equal but with equal and opposite phase angles (i.e. when $Z(\phi) = Z(-\phi)$ where Z is the magnitude of the impedances and ϕ is the phase angle). By employing equal and opposite phase angles, the two parts of the circuit are brought into resonance to ensure the maximum transfer of power. This result explains why some industrial mains power inputs incorporate *power factor correction* capacitors.

Transformer losses

The types of power loss which a transformer can suffer are the following:

- I^2R losses caused by the resistance of the windings.
- Eddy-current and stray inductance losses caused by unwanted magnetic interactions.
- Hysteresis loss arising from the core material, (if a core is used).

Taking these in turn, I^2R (or joule) losses are those which are always incurred in any circuit when a current, steady or a.c., flows through a resistance. These losses can be reduced in a transformer by making the resistance of each winding as low as possible consistent with the correct number of turns and the size of the transformer.

Joule losses are generally insignificant in small transformers used at radio frequencies; but they will cause overheating of mains transformers, particularly if more than the rated current is drawn or if ventilation is inadequate.

Stray inductance and eddy current losses are often more serious. An ideal transformer would be constructed so that all the magnetic field of the primary circuit coupled perfectly into the secondary winding. Only toroidal (ring-shaped) transformers come close to this ideal. In practice, the primary winding generates a strong alternating field which is detectable at some distance from the transformer, causing a loss of energy by what is termed *stray inductance*.

In addition, the alternating field of the primary can cause stray voltages to be induced in any conducting material used in the core or casing of the transformer, so that unwanted currents, called *eddy currents* flow. Since additional primary current must flow to sustain these eddy currents, they cause a loss of power which can be significant.

The problem of eddy currents in the core is tackled in two ways:

1. The core is constructed of thin laminations clamped together, with an insulating film coating on each to lessen or eliminate conductivity,
2. the core is constructed from a material that has high resistivity, such as ferrite.

The third type of loss, called *hysteresis loss*, occurs only when a magnetic core is used. It represents the amount of energy which is lost when a material is magnetized and de-magnetized. This type of loss can be minimized only by careful choice of the core material for any particular transformer.

Hysteresis losses will, however, increase greatly if the magnetic properties of the core material change, or if the material becomes magnetically saturated. The following precautions should therefore be taken in connection with transformers:

- Do not dismantle transformer cores unnecessarily, nor loosen their clamping screws.
- Never bring strong magnets near to a transformer core.
- Never pass d.c. through a transformer winding unless the rated value of the d.c. is known and is checked to be correct.

Specific transformer features
Mains supply

Mains frequency is low and fixed at either 50 or 60 Hz. A substantial core is required which must be laminated (hysteresis losses can be reduced to negligible proportions by careful choice of a core material). Where an external magnetic field is especially undesirable (as in audio amplifiers and cathode ray oscilloscopes), a toroidal core can be used with advantage. Figure 2.4(b) shows a mains type transformer with multiple tapping points to provide a range of output voltages.

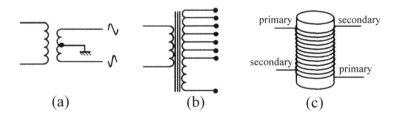

(a)   (b)   (c)

Figure 2.4   Types of transformer windings

Audio-frequency range

The core material must be chosen from materials causing only low hysteresis loss because of the higher frequencies that will be encountered. The windings must be arranged so that stray capacitance between turns is minimized. In general, any flow of d.c. is undesirable.

Lower-range radio frequency

In this range, the losses from laminated cores are unacceptably high, so that ferrite cores must be used. Because of the high frequencies involved, a small number of turns is sufficient for each winding. Stray fields are difficult to control, so that screening (see below) is often needed.

Higher-range radio frequency

Only air cores can be used in this range, and 'coils' may actually consist of less than one full turn of wire. They may even consist of short lengths of parallel wire. Unwanted coupling becomes a major problem, so that the physical layout of components near the transformer assumes great importance. Figure 2.4(a) shows how a RF circuit may be tapped to provide anti-phase outputs.

Standard windings

Centre-tapped secondary windings, Figure 2.4(a), are a good way of obtaining phase-inverted signals for audio amplification or for rectification purposes. This principle can be extended to the auto-transformer, which is a single-tapped winding equivalent to the use of a double-wound transformer with one end of the primary connected to one end of the secondary. The ratio of input/output voltages and currents still follows the normal transformer relationships. An auto-transformer with a variable tapping position (such as the Variac™) is used for providing variable voltage a.c. supplies. Note, however, that such transformers provide no isolation between their primary and secondary windings. Figure 2.5(b) shows a multi-tapped winding to provide a range of voltages.

Bifilar winding

See Figure 2.4(c). This is a method of providing very close coupling between primary and secondary windings, particularly useful in audio transformers. In this method of construction, the primary and the secondary turns are wound together, rather than in separate layers.

Shielding or screening

Components such as transistors, wiring and inductors may need to be shielded from the radiation created by transformers. Electrostatic screening is comparatively easy, for any earthed metal will screen a component from the electrostatic field of a transformer (though with very high frequencies a metal box which is almost watertight may have to be used). However, interaction between multi-windings needs to be prevented by placing an earthed metallic shield between the windings.

Electromagnetic screening, on the other hand, calls for the use of high permeability alloys such as mu-metal, or super-Permalloy to encase the device. Boxing a component in a such a way ensures that any magnetic fields are contained within the casing thus preventing the fields from escaping into the space outside.

Activity 2.2

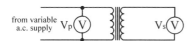

Using a transformer of known turns ratio, preferably a type using a tapped secondary winding, connect the circuit shown, left. Measure the a.c. input and output voltages for each set of taps, and find the values of V_S/V_P. Compare these values with the known values of the turns ratio.

An effective component to use in this experiment is a toroidal core with a 240 V primary winding, obtainable from most educational suppliers.

Transformer faults

The following are common transformer faults, with hints on how to detect and remedy them:

1. Open-circuit windings, can be detected by ohmmeter tests. A winding may also acquire high resistance which is caused by high resistance internal connections, typically 100K instead of 100 ohms.

2. Short-circuit turns, which are difficult to detect because the change of resistance is very small. Even a single turn that is short-circuited will dissipate considerable energy whilst making practically no difference to the d.c. resistance, thus making such very difficult to locate. S/c turns will cause an abnormally large primary current to flow when the secondary is disconnected, so that mains transformers overheat and transformers operating at high frequencies fail completely. This is a fault which particularly affects TV line output transformers. The simplest test and cure is replacement by a component known to be good.

3. Loose, damaged or missing cores. Loose cores will cause mains transformers to buzz and overheat. Cracked or absent ferrite cores in radio-frequency transformers will cause mistuning of the stage in which the fault occurs.

Thermal protection of transformers

Power supply circuits almost invariably contain semiconductor devices such as triacs (see Chapter 3) to provide additional protection for the main stages of the equipment, but for transformer protection the following further devices are used.

Fuses

Fuses or fuse-links are the most commonly used safety critical device for protection against overload conditions, but they are also open to mis-use. The fuse link should always be replaced with one of the correct rating after the fault condition has been cleared.

The fuse link rating is the current that the device will carry for a long period of time. It will often carry a current that is 25% in excess of the rated value for 1 hour or more. A further parameter is the Joule rating which is given by I^2t showing that the failure depends upon the square of the current and the time for which it is flowing. The minimum fusing current is typically 50 to 100% above the rated value. Hence many fuse links are described as *slow-blow* devices.

Fuse links also have a voltage rating that is different for a.c. and d.c. applications.

Embedded PTC (positive temperature coefficient) thermistors

.These devices are often buried within the winding space of a transformer and wired in series with the primary winding. When the current exceeds some value the temperature of the transformer rises, increasing the resistance of the thermistor so that the input current falls to lower the temperature.

Bimetallic switch

An embedded bimetallic operated switch with its contact wired in series with the primary current has been used in the past. The switch contacts open when the transformer temperature rises above some predetermined level to provide protection. When the temperature falls this system is self resetting.

- Note: for all embedded devices, a continuous cyclic switching action indicates a fault in urgent need of attention.

Answers to questions

2.1   60 V.

2.2   0.25 A.

2.3   5:1 stepdown.

3

Transistors and other semiconductor devices

The atom

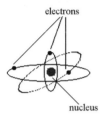

electrons

nucleus

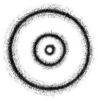

Germanium and silicon

A useful picture of an atom visualises it as a positively charged central core (the nucleus) surrounded by electrons. These electrons are arranged in bands, with a value of energy that is determined by which band an electron is in. The electrons in the lowest (outermost) energy band are most easily removed, and this band is called the *valency band*. Valency is a chemical idea that expresses how many atoms of one type will chemically combine with another type of atom, because chemical combination is the result of atoms exchanging or sharing their valency electrons. Alloys of metals also involve sharing of valency band electrons. The physical importance of electrons in the valency band is that they are mainly responsible for conduction of electricity in solid materials. A later vizualisation of the atom represents these bands of electrons as fuzzy spheres, rather than strictly drawn orbits.

Before studying the relative merits of diodes and transistors it is instructive to consider some of the features associated with the elements silicon and germanium.

Germanium (chemical symbol Ge) is a relatively rare element which has 32 electrons in its atomic structure with 4 in the valency band. By comparison, silicon (Si) which is found as silicon oxide (sand) is the second most abundant element. Silicon has 14 electrons in its atomic structure, again with 4 electrons in the valency band. Thus in the smaller silicon atom the valency electrons are more tightly bonded to the nucleus than those in germanium.

Although germanium has a lower inherent threshold voltage (100 mV) than silicon, it now finds relatively few applications in modern semiconductor technology. For both germanium and silicon, the reverse biased leakage current of a PN junction, approximately doubles for every 11°C rise in temperature. Since germanium has a significantly higher leakage at the same temperature, its operating range is restricted to about 75°C maximum. This compares unfavourably with silicon which can operate effectively up to around 250°C. The PN junction threshold voltage for silicon is about 600 mV.

Due to these advantages, silicon devices tend to dominate the semiconductor market place. Most recently in the search for operation at higher frequencies and lower power consumption and noise, an alloy of germanium and silicon (SiGe) has been introduced into semiconductor production technology, and compound semiconductor materials such as gallium arsenide are used for specialized purposes.

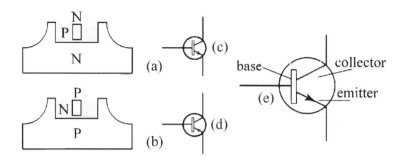

Figure 3.1 The bipolar transistor

The semiconductor materials (Ge and Si) that are used to make diodes and transistors are first purified to a very high degree and then in order to be able to control the conduction through the devices, impurities are added. These impurity elements are chosen from certain trivalent (3) elements such as aluminium (Al), gallium (Ga), and Indium (In) for P type and pentavalent (5) elements such as phosphorus (P), arsenic (As) and antimony (Sb). The impurity elements donate electrons (pentavalent) and holes (trivalent) to the conduction process.

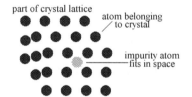

Holes exist only within a crystal that contains an impurity atom. When this atom has fewer valency electrons than the atoms around it, the place where a valency electron should have fitted is now a hole, and such spaces in a crystal structure can behave like a positively charged particle within the atomic structure. A hole within a crystal can behave as if it had mass and charge, and it can also move, but it cannot exist outside the crystal. You can form beams of electrons in a vacuum, but you cannot have beams of holes.

Bipolar junction transistors (BJTs) each have two junctions and three separate connections, as shown in Figure 3.1. The NPN transistor (a) has a thin layer of P-type material sandwiched between thicker N-type layers, whilst the PNP transistor (b) has a thin layer of N-type material sandwiched between thicker P-type layers.

The layer which forms the middle of the sandwich is called the base; the other two are called the emitter and the collector respectively.

Figures 3.1(c) and 3.1(d) illustrate schematically the NPN and the PNP transistor, respectively. The identity of the three connections is shown in Figure 3.1(e). The direction of the arrow-head on the emitter symbol distinguishes the transistor illustrated as being the NPN type. (The arrowhead points in the conventional + to – direction of current flow, as for the arrowhead of a diode symbol.) By comparison, diodes are formed with a single PN junction and just two connections.

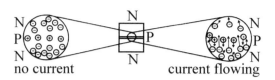

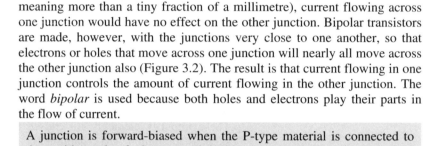

Figure 3.2 Current flow in bipolar transistor

BJT action

If the two junctions in any transistor were far apart ('far' in this case meaning more than a tiny fraction of a millimetre), current flowing across one junction would have no effect on the other junction. Bipolar transistors are made, however, with the junctions very close to one another, so that electrons or holes that move across one junction will nearly all move across the other junction also (Figure 3.2). The result is that current flowing in one junction controls the amount of current flowing in the other junction. The word *bipolar* is used because both holes and electrons play their parts in the flow of current.

A junction is forward-biased when the P-type material is connected to the positive pole of a battery and the N-type material is connected to the negative pole of the same battery of supply.

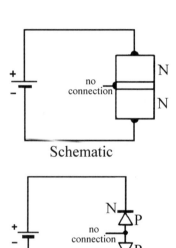

Consider an NPN transistor connected as in the upper diagram, left. With no bias voltage, or with reverse bias, between the base and the emitter connections, there are no carriers in the base-emitter junction, and the voltage between the collector and the base makes this junction reverse-biased, so that no current can flow in this junction either. The transistor behaves as if it were two diodes connected anode-to-anode (lower diagram). No current could flow in the circuit even if the battery connections were to be reversed.

When the base-emitter junction is forward-biased, however, electrons will move across this junction. Because the collector–base junction is physically so close, the collector potential will cause most of the electrons to be swept across this junction to provide collector current even if the junction is reverse-biased.

With both junctions conducting, most of the current will flow between the collector and the emitter, since this is the path of lower resistance. The transistor no longer behaves like two back-to-back diodes because the electrons passing through the base-emitter junction make the collector–base junction conduct despite the reverse bias between collector and base. (Note, that semiconductor diodes can be tested using this technique.)

The current flowing between the collector and the emitter is much greater (typically 25 to 800 times greater) than the current flowing between the base and the emitter. If the base is now unbiased or reverse-biased again, no current can flow between the collector and the emitter. Thus the current in the base-emitter junction controls the amount of current passing through the collector–base junction.

Activity 3.1

Using a silicon NPN transistor such as either 2N3053, 2N1711, 2N2219 or BFY50, measure the resistance between leads, using a multimeter set on the ohms scale. Remember that the multimeter polarity (+ and −) markings are reversed when the ohms scale is used, so that the terminal marked (+) is negative for ohms readings and the terminal marked (−) is positive.

Set out your readings as in the table below:

R_{be}	R_{be}	R_{bc}	R_{bc}	R_{ce}	R_{ce}
b+	b−	b+	b−	b+	b−

Note that each reading is taken in both directions so that, for example, R_{be} (b+) means a connection between base and emitter with the base (+), and R_{be}(b−) means the same connections but with the polarity reversed.

Note that if the connections to the transistor are unknown, the base lead can be identified since only the base will conduct to the other two electrodes with the same polarity.

When an NPN transistor is tested, the base will conduct to either emitter or collector when the base is positive. The base of the PNP transistor will conduct to both emitter and collector when the base is made negative.

Repeat the tests, and fill in a new table using a PNP transistor such as the 2N2905 or BFX39.

Now try to identify the leads of an unmarked transistor.

Ohmmeter tests

Tests with an ohmmeter can identify junction faults. A good transistor should have a very high resistance reading between collector and emitter with either polarity of connection. Measurements between the base and either of the other two electrodes should show one conducting direction and one non-conducting direction. Any variation from this pattern indicates a faulty transistor with either an open circuit junction (no conduction in either direction) or excessive leakage (conduction in both directions).

Current gain

The amount of current flowing between the collector and the emitter of a bipolar transistor is much greater than the amount of current flowing between the base and the emitter, but that collector current is controlled by the base current. The ratio, *collector current/base current* is in fact constant (given a constant collector-to-emitter voltage), and is commonly called the current gain for the transistor (its full name is the common-emitter current gain). The symbol used to indicate it is h_{fe}. A low-gain transistor might have a value of h_{fe} of around 20 to 50, a high-gain transistor one of 300 to 800 or even more.

Note that the tolerance of values of h_{fe} is very large, so that transistors of the same type, even transistors coming from the same batch, may have widely different h_{fe} values.

Activity 3.2

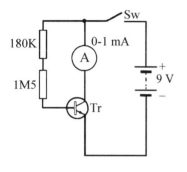

Measure the h_{fe} values for a number of transistors, using a transistor tester.

If a tester is not available, use the circuit, left, which will give approximate h_{fe} values for a silicon NPN transistor by the current readings on the multimeter when compared with the figures in Table 3.1.

Table 3.1   Meter readings and h_{fe} values

Meter reading	h_{fe}	Meter reading	h_{fe}
1 mA	200	0.5 mA	100
0.9 mA	180	0.4 mA	80
0.8 mA	160	0.3 mA	60
0.7 mA	150	0.2 mA	40
0.6 mA	120	0.1 mA	20

These results are achieved because the two base resistors maintain current flow at about 5 μA.

Question 3.1

A transistor has a stated value of h_{fe} equal to 150. What base current (in microamps) would you expect to produce a collector current of 60 mA?

Characteristics

Characteristics are the graphs that illustrate the behaviour of the transistor. The characteristics of a typical silicon transistor are shown in Figure 3.3, opposite.

Figure 3.3(a) shows the input characteristic, or the I_{be}/V_{be} graph. The slope of the line on this graph gives the inverse of the input resistance of the transistor, and its steepness shows that the input resistance is small. The fact that the graph line is curved shows that input resistance varies according to the amount of current flowing, and is greatest when the current flow is small. Figure 3.3(b) shows the I_{ce}/I_{be} response, called the transfer characteristic. This graph is a nearly straight line whose slope is equal to the current gain, h_{fe}.

Figure 3.3(c) shows the output characteristic, I_{ce}/V_{ce}, whose slope gives the value of output resistance. The horizontal parts of the graph lines show that a change in collector voltage has almost no effect on collector current flow. It is as if the transistor output had a resistance of very high value in series with it.

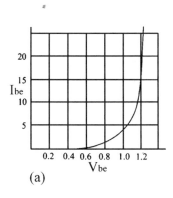

(a)

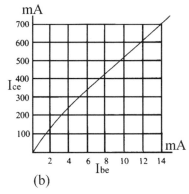

(b)

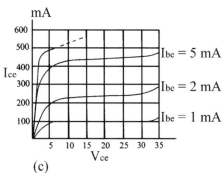

(c)

Figure 3.3   NPN silicon transistor characteristics

These graphs show that a transistor connected with its emitter common to both input and output circuits has a low input resistance, fairly large current gain and high output resistance.

Another graph which is very useful is the mutual characteristic, I_{ce}/V_{be} (or g_m, called the *mutual conductance*) shown here, left, for a typical power transistor. Note the large current values and the nearly straight line of the characteristic.

Although the I_{ce}/V_{be} is often a useful characteristic for an amplifier designer to know, it is not always provided by the transistor manufacturer. The value, however, does not need to be provided because it depends more on the amount of steady collector current, as set by bias, than anything else. At normal temperatures, around 25°C, the value of g_m for a silicon transistor is $40 \times I_c$ mA/V where I_c is the steady collector current, with no signal applied, per milliamp of collector current. In other words, for 1 mA collector current g_m is 40 mA/V, for 2 mA collector current, g_m is 80 mA/V and so on.

The quantities g_m, h_{fe} and r_{be} (resistance, base to emitter) are simply related by the formula:

$$r_{be} = h_{fe}/g_m$$

which gives r_{be} in units of K if g_m is in its normal units of mA/V. For example, if a transistor has the h_{fe} value of 100 and is used at a bias current of 1 mA, then its g_m value is 40 mA/V, and the r_{be} value is 100/40 = 2K5.

Question 3.2

A silicon transistor is being used with a steady collector current bias of 2.5 mA. What value of g_m would you expect?

Rules for substitution

When transistors are substituted one for another, the following rules should be obeyed:

- The substitute transistor must be of the same type (i.e., silicon, NPN, switching as opposed to amplifying, etc.).
- The substitute transistor should have about the same h_{fe} value.
- The substitute transistor should have the same ratings of maximum voltage and current.
- When making such substitutions, it is not always possible to guarantee the in-circuit performance of the change.

Applications of bipolar transistors (BJT)

Bipolar transistors are used as current amplifiers, voltage amplifiers, oscillators and switches. An amplifier has two input and two output terminals, but a transistor has only three electrodes. It can therefore only operate as an amplifier if one of its three electrodes is made common to both input and output circuits.

Any one of a transistor's three electrodes can be connected to perform in this common role, so there are three possible configurations: *common-emitter*, *common-collector* and *common-base*. The three types of connection are shown in Figure 3.4(a), (b) and (c) respectively.

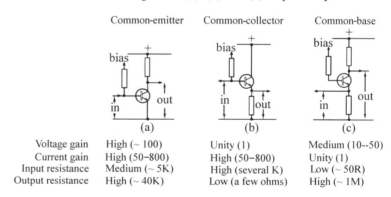

	Common-emitter	Common-collector	Common-base
Voltage gain	High (~ 100)	Unity (1)	Medium (10--50)
Current gain	High (50–800)	High (50–800)	Unity (1)
Input resistance	Medium ($\sim 5K$)	High (several K)	Low ($\sim 50R$)
Output resistance	High ($\sim 40K$)	Low (a few ohms)	High ($\sim 1M$)

Figure 3.4   The three circuit connections of a bipolar transistor

The normal function of a transistor when the base-emitter junction is forward biased and the base-collector junction reverse-biased, is as a current amplifier. Voltage amplification is achieved by connecting a load resistor (or impedance) between the collector lead and the supply voltage, see Figure 3.4(a). Oscillation is achieved when the transistor is connected as an amplifier with its output fed back, in phase, to its input. The transistor can also be used as a switch or relay when the base-emitter junction is switched between reverse bias and forward bias.

The three basic bipolar transistor circuit connections are shown in Figure 3.4, with applications and values of typical input and output resistances given below each. Figure 3.4(a) shows the normal amplifying connection used in most transistor circuits. The common-collector connection in Figure 3.4(b), with signal into the base and out from the emitter, is used for matching impedances, since it has a high input impedance and a low output impedance. The common-base connection, with signal into the emitter and out from the collector, shown in Figure 3.4(c) is nowadays used mainly for UHF amplification.

Transistor failure

When transistors fail, the fault is either a short-circuit (s/c) or an open-circuit (o/c) junction; or the failure may possibly be in both junctions at the same time. An o/c base-emitter junction makes the transistor 'dead', with no current flowing in either the base or the collector circuits.

When a base-emitter junction goes o/c, the voltage between the base and emitter may rise higher than the normal 0.6 V (silicon) or 0.2 V (germanium) though higher voltage readings are common on fully operational power transistors when large currents are flowing. An s/c base-emitter junction will allow current to flow easily between these terminals with no voltage drop, but with no current flowing in the collector circuit.

The two above faults are by far the most common, but sometimes a base-collector junction goes s/c, causing current to flow uncontrollably. The base region of a bi-polar device can be ruptured through the application of an excessively high collector voltage, this is often described as *punch-through*.

All of these faults can be found by voltage readings in a circuit, or by use of the ohmmeter or transistor tester when the transistor is removed from the circuit. (Transistor testers are now available that allow in-circuit testing.)

Field-effect transistors

To be strictly accurate, the so-called *field effect transistor* is not really a transistor at all. The word '**transistor**' is a compression of the term '*transfer resistor*', and the FET (as it is commonly abbreviated) does not work like that at all. It is, in fact, more correctly described as a field-effect device, because its operation depends on the presence and effects of an electric field. Nevertheless, the misnomer 'field-effect transistor' has become well established and must be used here.

The bipolar transistor relies for its action on making a reverse-biased junction conductive by injecting current carriers (electrons or holes) into it from the other junction. The principles of the field-effect transistor (FET) are entirely different. In any type of FET, a strip of semiconductor material of one type (P or N) is made either more or less conductive because of the presence of an electric field pushing carriers into the semiconductor or pulling them away.

There are two types of field-effect transistor, the junction FET and the metal-oxide-silicon FET, or MOSFET. Both work by controlling the flow of current carriers in a narrow channel of silicon. The main difference between them lies in the method used to control the flow.

Junction FET

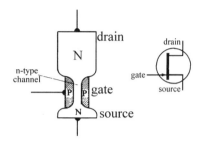

Look first at the structure of the junction FET. A tiny bar of silicon of either type (the N-type is illustrated) has a junction formed near one end. Connections are made to each end of the bar, and also to the P-type material at the junction. The P-type connection is called the *gate*, the end of the bar nearest the gate is the *source*, and the other end of the bar is the *drain*. A junction FET of the type illustrated is normally used with the junction reverse-biased, so that few moving carriers are present in the neighbourhood of the junction. This way of using a JFET is also termed *depletion mode*.

The junction, however, forms part of the silicon bar, so that if there are few carriers present around the junction, the bar itself will be a poor conductor. With less reverse bias on the junction, a few more carriers will enter the junction and the silicon bar will conduct better; and so on as the amount of reverse bias on the junction decreases.

When the voltage is connected between the source and the drain, therefore, the amount of current flowing between them depends on the amount of reverse bias on the gate; and the ratio is called the mutual conductance, whose symbol is g_m. This quantity, g_m, is a measure of the effectiveness of the FET as an amplifier of current flow.

For most FETs, g_m values are very low, only about 1.2 to 3 mA/V, as compared with corresponding values for a bipolar transistor of from 40 mA/V (at 1 mA current) to several amperes/volt at high levels of current flow. Because the gate is reverse-biased, however, practically no gate current flows, so that the resistance between gate and source is high; very much higher than the resistance between base and emitter of a working bipolar transistor.

MOSFET

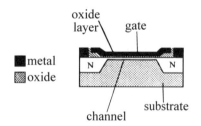

The drawing, below left, shows the basic construction of the metal-oxide-silicon FET or MOSFET. A silicon layer, called the *substrate* or *base*, is used as a foundation on which the FET is constructed. The substrate may have a separate electrical connection, but it takes no part in the FET action and if a separate electrical connection is provided it is usually connected either to the source or the drain. Two regions which are both doped in the opposite polarity to the substrate are then laid on the substrate and joined by a thin channel. In the illustration the substrate is of P-type silicon and the source, drain and channel are of N-type so that there is a conducting path between the N-type source and drain regions.

The gate is insulated from the channel by a thin film of silicon oxide, obtained by oxidizing some of the silicon of the channel, and a metal film is deposited over this insulating layer to form the gate itself.

A positive voltage applied to the gate has the effect of attracting more electrons into the channel, and so increasing its conductivity. A negative potential so applied would repel electrons from the channel and so reduce its conductivity.

Both N-channel and P-channel devices can be made. In addition, the channel can be either doped or undoped (or very lightly doped). If the channel is strongly doped there will be a conducting path of fairly low resistance between the source and the drain when no bias is applied to the gate. Such a device is usually operated with a bias on the gate that will reduce the source-drain current, and is said to be used in *depletion mode*. When the channel is formed from lightly-doped or undoped material it is normally non-conducting, and its conductivity is increased by applying bias to the gate in the correct polarity, using the FET in *enhancement mode*. Enhancement mode is more common.

With the gate-to-source voltage equal to zero, the device is cut off. When a gate voltage that is positive with respect to the channel is applied, an electric field is set up that attracts electrons towards the oxide layer. These now form an induced channel to support a current flow. An increase in this positive gate voltage will cause the drain-to-source current flow to rise.

FET handling problems

Junction FETs cause few handling problems provided that the maximum rated voltages and currents are not exceeded. MOSFETs, on the other hand, need to be handled with great care because the gate must be completely insulated from the other two electrodes by the thin film of silicon oxide. This insulation will break down at a voltage of 20 to 100 V, depending on the thickness of the oxide film. When it does break down, the transistor is destroyed.

Any insulating material which has rubbed against another material can carry voltages of many thousands of volts; and lesser electrostatic voltages are often present on human fingers. There is also the danger of induced voltages from the a.c. mains supply.

Voltages of this type cause no damage to bipolar transistors or junction FETs because these devices have enough leakage resistance to discharge the voltage harmlessly. The high resistance of the MOSFET gate, however, ensures that electrostatic voltages cannot be discharged in this way, so that damage to the gate of a MOSFET is always possible.

To avoid such damage, all MOS gates that are connected to external pins are protected by diodes which are created as part of the FET during manufacture and which have a relatively low reverse breakdown voltage. These protecting diodes will conduct if a voltage at a gate terminal becomes too high or too low compared to the source or drain voltage level, so avoiding breakdown of the insulation of the gate.

The use of protective diodes makes the risk of electrostatic damage very slight for modern MOS devices, and there is never any risk of damage to a gate that is connected through a resistor to a source or drain unless excessive d.c. or signal voltages are applied. Nevertheless, it is advisable to take precautions against electrostatic damage, particularly in dry conditions and in places where artificial fibres and plastics are used extensively. These precautions are:

- Always keep new MOSFETs with conductive plastic foam wrapped round their leads until after they have been soldered in place.

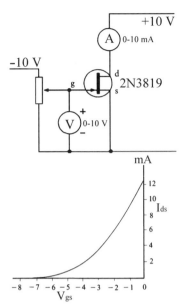

FET applications

Activity 3.3

Using a 2N3819 junction FET, connect the circuit shown, above left. The milliammeter measures the current through the channel, and the voltmeter measures the (negative) bias on the gate. For bias voltages from zero to 9 V, record the drain currents and plot the characteristic curve which should look similar to that illustrated. Record also the cut-off voltage.

FETs can be used in circuits similar to those in which bipolar transistors are used, but they give low voltage gain and are only used when their peculiar advantages are required.

These are:

• FETs have a very high input resistance at the gate, a useful feature in voltmeter amplifiers.

• FETs perform very well as switches, with channel resistance switching between a few hundred ohms and several megohms as gate voltage is varied.

• The graph of channel current Ids, plotted against Vgs, the voltage between the gate and the source is noticeably curved in a shape called a square law (see the graph obtained in Activity 3.3). This type of characteristic is particularly useful for signal mixers in superheterodyne receivers.

Double-gate MOSFETs are used as mixers and as RF amplifiers in FM receivers. The shape of their characteristic also gives less distortion in power amplifiers, and high-power FETs are available for use in high-quality audio equipment

Question 3.3

Some military applications make use of thermionic valves for circuits that could just as well be performed by transistors. Why?

FET and MOSFET failure

Failure of a junction FET can be caused by either an open-circuit or a short-circuit junction. MOSFET failure is almost always caused by breakdown of the insulating silicon oxide layer. In either case, gate voltage can no longer control current flow in the channel between source and drain, and pinch-off (the cut-off condition) becomes impossible.

If very large currents have been allowed to flow between source and drain, the channel may overheat and rupture.

Thyristor-type devices

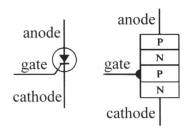

This topic covers a range of related semiconductor devices that are basically 4 layer PNPN sandwiches designed either for a.c. or d.c. switching purposes. The sketch, left, shows the symbol and the construction.

Silicon controlled rectifier (SCR). This is a medium to high power diode with a gate electrode fabricated on the cathode section. The SCR is designed as a fast acting voltage controlled relay for d.c. switching and is self latching in operation.

With the anode voltage positive with respect to (*wrt*) the cathode, the action is controlled by the gate voltage. Briefly applying a positive voltage to the gate (V_g) causes the SCR to go into conduction and then remain in this state even if V_g is removed. Turn off is achieved by briefly reducing the anode current below some threshold value, typically a few milliamps. The device can carry several tens of amps at several hundreds of volts.

Silicon controlled switch (SCS). This device has a similar construction to the SCR, is self latching and functions in a similar way for turn-on, but is turned off by either reducing the anode current or reverse biasing the gate to cathode junction.

Thyristor. The thyristor which is a switching device, and often referred to as silicon-controlled rectifier. It is fabricated as four doped layers and its symbol and schematic construction are as shown above left.

When gate voltage is zero, the device offers high resistance irrespective of the polarity of the anode-to-cathode voltage, because there is always at least one PN junction reverse-biased. Even when a thyristor is forward-biased, with the anode positive with respect to the cathode, it will only be driven into conduction when a short positive pulse is applied to its gate.

Once it has been switched on in this way, the thyristor will remain conducting until either:

- the voltage between anode and cathode falls to a small fraction of a volt; or

- the current flow between anode and cathode falls to a very low value.

The important ratings for a thyristor are its maximum average current flow, its peak inverse voltage, and its values of gate-firing voltage and current. A small thyristor will fire at a very small value of gate current, but a large one calls for considerably greater firing current (100 µA or more).

Question 3.4

A power controller for an electric drill has failed, allowing the drill to run at maximum speed but with no speed control. Can you suggest the most likely reason for this failure?

Activity 3.4

Connect up the circuit illustrated, left. Any small 1 Amp thyristor such as BTX 18-400, TICP106D will be suitable. The purpose of the lamp bulb is to indicate when the thyristor has switched on.

Use a d.c. supply as shown, and measure voltage and current at the gate just as the thyristor fires noting also that both values change after the thyristor has fired. After the lamp lights, disconnect the gate circuit, and note that this has no effect. Switch off the power supply, and then switch on again. The lamp will not now light because the thyristor has switched off.

Repeat the operation, this time using an unsmoothed supply. The thyristor now switches off when the gate voltage is switched off because the unsmoothed supply reaches zero voltage 100 times per second (assuming a full-wave rectified 50 Hz supply).

Thyristors are used for power control, using either a.c. or an unsmoothed rectified supply. By being made to fire at different points in each a.c. cycle, the thyristor can be made to conduct for different percentages of the cycle, thus controlling the average current flow through the load. This is called *phase control* of the duty cycle.

An alternative method used for slow changing load conditions such as in furnace temperature control, fires the thyristor into conduction for several a.c. cycles and then drives into cut-off for several further cycles. The average power in the load can be controlled by altering the ratio of *conducting cycles* to *non-conducting cycles*. This technique is known as *burst control*.

See Chapter 10 for a more complete explanation of the action of a thyristor in these applications.

Gate-turn-off thyristor (GTO). This is an alternative device which is triggered on by a positive pulse of fairly high current (typically 100 mA) and triggered off by a negative pulse of lower current; both pulses being applied to the same gate.

Tetrode thyristor. This variant is equipped with anode and cathode gates to provide a self latching action. The turn-on action is achieved by driving

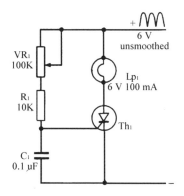

Thyristor failure

The diac

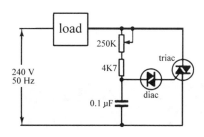

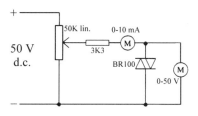

the cathode gate positive or the anode gate negative, whilst the device is turned off by reversing the gate potentials.

Activity 3.5

Connect up the thyristor circuit illustrated, left, and switch on. Use the potentiometer to control lamp brightness by altering the time in the cycle at which the thyristor fires. Note that an unsmoothed full-wave rectified supply is essential.

Failure of a thyristor can be caused by an open-circuit gate, or by internal short-circuits. When the gate is open-circuit, the thyristor will fail to conduct at any gate voltage. When an internal short circuit is present, the thyristor acts like a diode, conducting whenever the anode is more than 0.6V positive *wrt* the cathode, so making control impossible. A completely short-circuit thyristor is able to conduct in either direction, with similar total loss of control.

Most thyristor circuits include a low pass filter consisting of an inductor and capacitor (called a *snubber* circuit) in the anode circuit, in order to suppress the radio-frequency interference which is caused by the sudden switch-on action of the thyristor.

A diac is a two terminal trigger device fabricated as an inverse coupled pair of diodes and is often used in the gate circuit of a thyristor or triac, see the drawing, left. At low voltages or either polarity, a diac is completely non-conducting. However, it can be driven into conduction with a triggering voltage of either polarity.

When a diac is used, the gate of the thyristor can be connected to the cathode by a low-value resistor to avoid accidental triggering. In addition, if the triggering waveform rises slowly, the diac will ensure that the thyristor is switched on by a fast rising pulse of current, so avoiding uncertain triggering times. The symbol for, and a typical use of, the diac are shown here.

Activity 3.6

Connect a diac, such as the BR100, into the circuit illustrated here. Ensure that the potentiometer is set so as to provide zero voltage at the output when the power is applied, and then switch on. Read the values of V and I for each 5 V step of voltage until current starts to flow, and then on each 2 mA change of current up to 10 mA. Now return the potentiometer setting to the zero-output position and reverse the polarity of the power supply. Repeat the readings. Plot a graph of V against I, allowing for both positive and negative values.

The triac

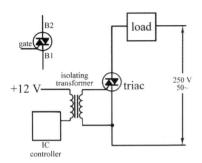

A triac is a two-way thyristor which, when triggered, will conduct in either direction. The terminals are labelled as B1, B2 and gate (the words 'anode' and 'cathode' cannot be used in this case because the current can flow in either direction). The gate can be triggered by a pulse of either polarity, but the most reliable triggering is achieved when the gate is pulsed positive with respect to B1.

In most triac circuits, B1 and B2 have alternating voltages applied to them, so that the gate must receive the same waveform as B1 when it is not being triggered. To trigger the gate, a pulse must be added to the waveform already present, and this is most easily done by using a pulse transformer driven by a trigger circuit. The transformer being used to isolate the triac, which works at line voltage, from the low-voltage control circuit. The drawing, left, shows the symbol for a triac, and part of a typical triggering circuit.

- The causes of failure in triacs are the same as they are for thyristors.

The advantages of using thyristors and triacs for power control are:

1. Both thyristor and triac are either switched completely off, with no current flowing, or fully on, with only a small voltage between the terminals. Either way, very little power is dissipated in the semiconductor, so that heat sinks are required only for devices that handle a substantial amount of power.

2. Operation can be either at line or at higher frequencies, unlike relays or similar electromechanical switches.

3. Thyristors and triacs are 'self-latching' which means that they stay conducting once they have been triggered. A relay, by contrast, needs a current passed continuously through its coil to keep it switched over, and relay latching circuits require additional relay contacts.

Activity 3.7

HAZARD: High voltages. This circuit must be supplied from an isolating transformer, and all live parts of the circuit must be covered to protect against accidental contact.

Figure 3.5 shows a circuit using a triac to control the power to a 40 W lamp. The potentiometer adjusts the charging current to the capacitor, so determining at what part in each half-wave the voltage across the diac will be enough to cause conduction and so fire the triac. Note how the voltage registered on the a.c. voltmeter varies as the potentiometer is adjusted.

For further work, inspect the waveforms in the circuit across the capacitor and on each side of the diac. An oscilloscope can be used since the circuit is fed from an isolating transformer, but remember that an oscilloscope must never be used to check waveforms in a triac or thyristor circuit which is directly connected to the mains.

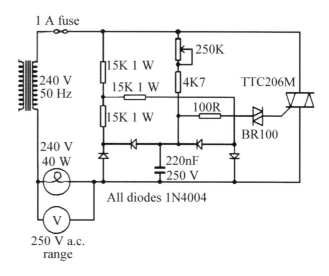

Figure 3.5 Circuit for Activity 3.7

Display devices
Vacuum fluorescent displays (VFD)

These devices are often used for the displays on CD type equipment, video recorders and audio systems. The operation of these devices is based on the same principles as thermionic valves and contain electrodes that function as cathode, grid and anode enclosed in a vacuum, but constructed as thin flat panels. The anode phosphors are deposited on the inside face of the front glass panel and are shaped to provide light emission that represents the desired function or legend. A full range of alpha-numeric characters is also included.

Typically a 30 V potential is applied between cathode and anode, but with the cathode at –30 V so that the anodes can be effectively grounded. This provides for a simpler driver stage which is coupled to the cathodes. As a power saving feature, the grids are pulsed sequentially but only the anode segments that are simultaneously pulsed will light.

For fault tracing purposes, the pulse trains can be monitored on an oscilloscope. For display failure, check the operating potentials (–30 V). If these are present, check the heater voltage supply (typically 4 V a.c.). Since the device is commonly mounted on a PCB, cracks in the tracks form a distinct possibility as the source of failure. Electrical leakage on the PCB due to spillage can cause spurious illumination of unwanted display segments. For a dim, low level display, check the 30 V supply. If this is low, check the quality of the power supply electrolytic capacitors.

Gas plasma display panels (GPDP)

A plasma is the region in an electrical discharge path in which the number of positive and negative ions is approximately equal so that the path is electrically neutral and highly conductive. The plasma is often described as the fourth state of matter (the other three being solid, liquid and gaseous).

Plasma display panels (PDP) may be manufactured as large, slim replacements for the colour television cathode ray tube. Their main features are high brightness, good colour and contrast scale, and large viewing angle. The lifetime tends to be a trade-off between brightness, contrast and power consumption. Displays capable of providing high definition television of up to 1024 lines at 1920 pixels per line have been manufactured.

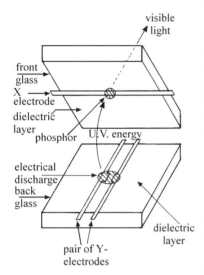

The PDP which is a.c. driven, is constructed around a pair of glass panels separated and sealed with a slim space to provide a set of chambers to hold a mixture of helium and xenon gases that provide the plasma. Electrode structures are electro-deposited on both glass plates at right angles to each other as shown in the single pixel arrangement shown, left, using the conductive properties of indium tin oxide (ITO). The pairs of Y electrodes are crenellated (crinkled) to encourage the plasma formation. The X electrodes deposited on the other glass plate supports red (R), green (G) and blue (B) phosphors with alternate rows carrying small areas of R, G and G, B phosphors on alternate strips. Thus two rows are needed to be energized simultaneously to produce groups of R, G, B pixels. The inner faces of both glass panels are electro-coated with a dielectric to form a barrier to prevent damage to the electrode structure through the action of the plasma. The plasma discharge generates ultra-violet (UV) light which activates the phosphors that in turn create visible coloured light.

Digital drive signals to the display panel controller provide 8 bits per RGB pixel to generate about 16.7 million possible different pixel colours.

Transducers

A transducer is a device that converts energy from one form to another. The transducers that are of most interest as far as electronics is concerned are those in which one of the energy forms is electrical energy. Of the many possible transducers, the most common are the transducers for light, heat and sound.

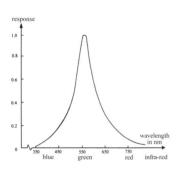

Photo-electric transducers are widely used in the communications and measurement environments. These devices convert light energy into electrical signals. They are used extensively in position sensing and counting devices where the interruption of a light beam is converted into an electrical signal. As most of these devices have a wavelength response greater than the human eye, their applications extend into the infra-red region. The drawing, left, shows the typical human-eye response in relationship to the infra-red and ultra-violet (less than about 340 nm) wavelengths.

Photo-conductive cells

The resistance of certain materials that are used for light sensitive devices decreases when subjected to increased illumination. The common materials used for light-dependent resistors (LDRs), include selenium, cadmium sulphide, cadmium selenide and lead sulphide. Each material responds to different wavelengths of light. Figure 3.6(a) shows the general appearance of a photo-conductive cell and its construction. Gold-conducting sections are deposited on a glass plate with a long, meandering gap to isolate the two sections. A thin layer of suitable photoconductive material is then deposited to bridge the insulating region. This construction is necessary to reduce the cell's resistance to a usable value.

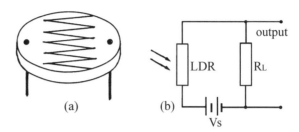

Figure 3.6   The photoconductive cell (a) and a typical circuit (b)

Typical resistance values for various cells range from about 500K to 10M in darkness and from about 1K to 100K in bright light. The change in resistance is non-linear and there is a significant time lag (up to 0.1 sec) in response to a pulse of light.

Figure 3.6(b) shows a basic circuit using a LDR. The current flowing in the load resistance R_L due to the supply voltage V_S produces the output voltage. Increasing the illumination of the LDR lowers its resistance so that the current, and hence the output voltage, increases. The maximum permitted voltage for these cells may be as high as 300 V and the maximum power dissipation is in the order of 300 mW. The dark-to-light resistance ratio ranges from about 50:1 to 250:1.

The cells are particularly sensitive to red light and the infra-red wavelengths, so that they are often used as flame detectors in boiler and furnace control systems. Application of an excessive voltage or current is the common cause of failure.

Question 3.5

A cadmium-sulphide LDR is often specified for oil-fired boiler flame detection because of its sensitivity to red light. Can you think of another good reason for using this type of component?

Photo-diodes

Photo-diodes are very similar to a normal PN junction device, they are formed with one very thin region and equipped with a lens so that light energy can be directed into the depletion region. When this happens hole-electron pairs are generated to increase the diode's conductivity. Such diodes have a peak response in the infra-red region, but the response to visible light is still very useful.

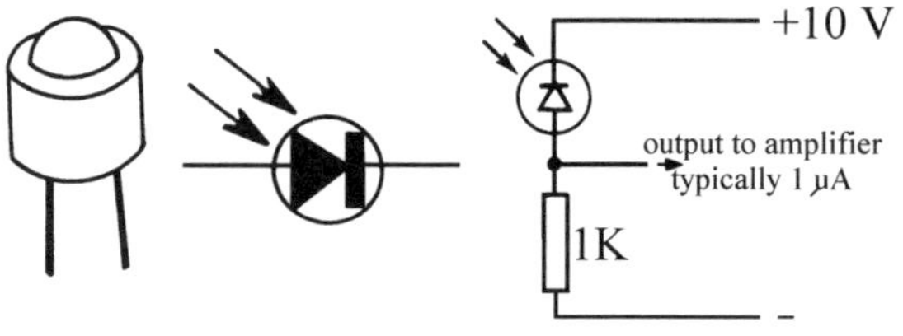

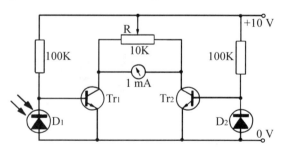

Figure 3.7 Photodiode and symbol (a) and typical circuit (b)

The diodes are operated either reverse-biased, or only very slightly forward-biased (to increase sensitivity) so that no current flows. Typical dark current is as low as 2 nA and rising to about 100 µA in bright light for germanium types. Modern PIN silicon versions using a PN junction with an additional interposed layer of I-type (intrinsic, meaning undoped) semiconductor may have a peak power dissipation greater than 100 mW.

The increase in current due to the light is practically linear. The response time to a pulse of light is very short so that they find applications in high-speed switching circuits. Generally, the small output is a disadvantage and amplification is needed. The basic circuit configuration is shown in Figure 3.7, above, whilst the more practical application of a light meter is shown in Figure 3.8, following.

With both diodes shielded from light, the conduction of the two transistors (Figure 3.8) is balanced by the potentiometer R so that there is no current flowing through the meter. In operation D_2 is maintained in darkness and the base voltage of Tr_1, and hence its collector current, depends on the level of light falling on D_1. An increase in the illumination of D_1 causes Tr_1 collector current to decrease, its collector voltage rises and so the meter deflects from its zero position. A meter of this type needs to be calibrated against a known reference light source.

Figure 3.8 Circuit of a typical light meter

Photo-transistors

A photo-transistor has a similar construction to a silicon planar transistor except that it is equipped with a lens so that light can be made to shine directly into the base-emitter junction. Their response time to a pulse of light is in the order of 1–2 µs and are therefore mostly used in medium-speed light detector circuits. The current ranges from about 25 to 50 μA in darkness up to about 5 to 10 mA in bright light.

The photo-transistor, Figure 3.9, is connected so that the base is either open-circuit, or reversed-biased, or only very slightly forward-biased. In some devices, the base connection is omitted altogether so that only the collector and emitter connections can be used. The load may be placed in either the collector or emitter leads as shown, left. Light shining into the base-emitter junction modulates the base current which is in turn magnified by the current gain. Such a device can be used to drive a load directly as shown in the relay-load example.

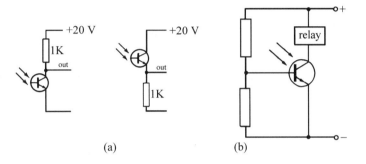

(a)   (b)

Figure 3.9 Photo-transistor use. (a) Load positions, (b) driving a relay (back-e.m.f. diode omitted)

Light emitting diodes (LEDs)

The operation of the photo-diode depends on the application of energy to generate hole-electron pairs. The reverse action is that when holes and electrons re-combine, energy is released. In germanium and silicon, this energy is released as heat into the crystal structure. However, in materials such as gallium arsenide and gallium phosphide this energy is released as light, and different semiconductor compounds release light of different wavelengths (colours).

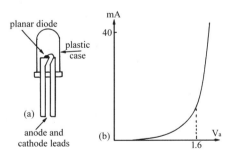

Figure 3.10 Construction of LED (a) and characteristic (b)

The basic structure of the diode is shown in Figure 3.10(a). The plastic moulding not only holds the component parts together but also acts as a light-pipe so that most of the light generated is radiated from the domed region.

A characteristic typical of a light-emitting diode is shown in Figure 3.10(b). When forward-biased beyond about 1.6 volts, the current rises rapidly and light is released. The current should be limited to a value less than about 40 mA by the use of a series resistor.

Opto-couplers

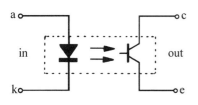

These consist of an LED and photo-transistor pair whose symbol is illustrated, left. The input signal modulates the diode current and hence the intensity of its light output. This variation in light produces a variation in collector current to provide an output signal. Since the light beam has no electrical impedance, there is no matching problems between input and output circuits. The electrical isolation is very high and an opto-coupler can withstand test voltages as high as 4 kV between input and output terminals. Alternatively the LED can be used to launch energy into a glass optical fibre cable to transmit the signal over very much greater distances. The light path within the opto-coupler can be interrupted by a shutter. This generates a pulsating signal that might be used to drive a counter circuit capable of working at high speed.

The photo-transistor may be replaced by a compound transistor (Darlington amplifier) to provide higher gain or by a photo-thyristor to provide higher output current. However both arrangements result in a lower switching speed.

The radiation from red LEDs contains a significant level of infra-red (IR) radiation. Similarly, many transistors show a good response to IR energy. Therefore selected devices can be used to provide an IR communications link.

Photo diodes and transistors have similar fault tolerance as conventional semiconductors. Excessive voltage, current or temperature are the common causes of failure. Faulty devices can be located using the same techniques as applied to other semiconductors.

Photovoltaic devices

A layer of selenium is deposited on an iron or aluminium backing plate which forms the positive pole. A transparent layer of gold is evaporated on to this to form the negative pole. A metallic contact ring completes the circuit. The general construction and circuit symbol is shown in Figure 3.11.

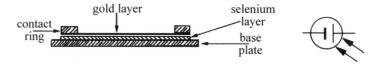

Figure 3.11   Section through photovoltaic cell, and circuit symbol

Light shines through the gold film into the layer of selenium. This releases electrons that form an electric field within the selenium, making the gold layer the negative pole of the cell. The whole cell is enclosed in a plastic housing with a transparent window for protection. The maximum current in bright light depends on the particular cell, but short-circuit currents in excess of 1 mA can be obtained. Such cells are suitable to drive a portable photographic light meter.

Activity 3.8

Using a video recorder or digital disk player equipped with a VFD device, measure and record the anode, cathode and grid voltages under various conditions of programme selection. Monitor the grid voltages using an oscilloscope during selection changes. Note and record the necessary electrode voltage polarities for correct operation of this thermionic device.

Solar cells

Solar cells are the semiconductor version of the selenium photovoltaic cell. They are formed from heavily doped PN junctions. A cell of about 4 cm^2 is capable of providing 0.6 V on open circuit, with a short-circuit current of up to 100 mA in bright light. The cells can be connected in a series-parallel configuration to provide a greater output power level. The cost of manufacturing such cells has fallen steadily, and they are used extensively to provide power for remote atmospheric observation stations.

Question 3.6

Since a photovoltaic cell can produce an output even on a fairly dull day, why do we not make more use of this to generate electricity?

The laser diode

Safety Note. All laser equipment should carry a permanent legible label warning of the dangers to unprotected eyes. In certain cases, the radiation is invisible. For safety reasons, protective safety spectacles that have an optical density of at least 6 for the appropriate wavelength should always be worn.

Light from conventional sources consists of a series of short bursts of waves that are all out of phase with each other, typically because each atom of a hot material is vibrating out of phase with others. The name laser is derived from the term *light amplification by the stimulated emission of radiation.* The device is basically an optically transparent cavity with a reflecting surface at one end and a partial reflector at the other. The stimulated emission is produced by the recombination of injected current carriers (holes and electrons) within the cavity.

This sets an oscillatory wave motion between the two ends and some of the energy emerges through the partial reflecting surface. When the forward current through the diode exceeds some threshold level, often as low as 20 mA for about 2.5 V of forward bias, the lasing action occurs, with all the emitted waves in phase. Increasing the current produces a linear increase in light output up to some saturation level. If the current is allowed to fall below the threshold level, then lasing ceases. Such laser diodes are often referred to as CW or continuous wave lasers.

There are many laser devices in use but those most commonly used for communications purposes are:

- Distributed feedback (DFB) lasers which contain an optical wavelength filter to control the wavelength of lasing.
- Vertical cavity surface emitting lasers (VCSEL) which is fabricated from ultra thin layers of Group III/V semiconductors (GaAs, AlAs, AlGaAs, and InAlGaP) and provide emission that is perpendicular to the flat upper surface.

The VCSEL type of construction provides for ease of packaging and coupling to optical fibres. These small flat devices are commonly used in CDROMs and similar replay machines. They provide about 5 mW of output power with a very narrow beam at about 640 nm. The threshold current of the VCSEL is about 30 mA.

Whilst both of these diode laser devices are fairly robust, they will suffer damage through excessive reverse current and high voltage transients.

Fibre optical links

Fibres made from specially produced silica glass can be made to function as light pipes. By converting the more normal electrical signals into pulses of light, these new signals can be transmitted over great distances with negligible losses. At the receiver, the light wave signals are converted back into electrical signals for normal communications purposes. Thus optical fibre communications links act somewhat like opto-couplers but with a very much longer reach. Because the glass fibre has no electrical impedance, there is no matching problem at either the sending or receiving ends. This feature also allows the light wave signals to pass through an electrically noisy environment without interference.

Figure 3.12 shows the basic construction of a length of optical fibre and its mode of signal transmission. The diagram shows the bi-cylindrical nature of this construction, with the core and cladding glasses having different optical densities. The outer cladding has a lower density than the core and this causes any light wave launched into the core to continually reflect from the interface between the two glasses. Any light signal launched into the fibre at one end will travel through to the far end with negligible losses. The system works best at infrared wavelengths between 1.35 and 1.55 μm, when the losses are typically around 0.15 dB/km. International networks are made up of many lengths of fibre joined together by fusion splicing and such joints can have a loss as small as 0.05 dB per connection.

Cheaper polymer (plastic) fibres can be constructed that function in a similar way but with a greater attenuation that restricts transmission lengths to about 100 metres.

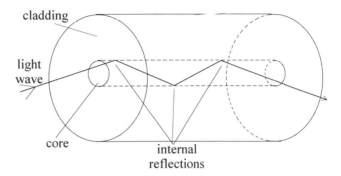

Figure 3.12   Construction of a typical optical fibre.

Activity 3.9

Using a TV receiver equipped with an IR remote control system, check and record the limiting distance of the control range. At about half of this range, set up a glass, metal and brick panel. From behind each panel in turn, attempt to control the TV receiver functions and record the results. Now transfer the IR remote control to the front of the 3 panels, aim the remote control at each in turn to check their reflective properties. This should provide an indication of how the local environment can influence the behaviour of the IR links that are being used as a means of extending the reach of local area networks.

Thermistors

These devices may be in rod, bead, washer or disc form. They are made from carefully controlled mixtures of certain metallic oxides. These are sintered at very high temperatures to produce a ceramic finish. Thermistors have a large temperature coefficient of resistance, given by the expression:

$$\frac{\text{change of resistance}}{\text{original resistance} \times \text{change of temperature}}$$

In normal resistors, this value is relatively small, typically in the order of 3×10^{-3}. For thermistors however, the temperature coefficient can be of the order of 15 to 50×10^{-3}. Negative temperature coefficient (NTC) thermistors have resistances that fall with a rise in temperature and are commonly made from mixtures of the oxides of manganese, cobalt, copper and nickel. Positive temperature coefficient (PTC) components whose resistance increases with a rise in temperature can be made from barium titanate with carefully controlled amounts of lead or strontium.

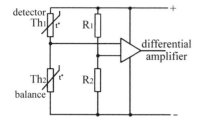

Thermistors are used extensively for temperature measurement and control up to about 400°C. A typical circuit is shown here, in which the bridge circuit is balanced at some low temperature so that the output from the differential amplifier is zero. Th_2 is maintained at this low temperature, whilst Th_1 is exposed to the temperature to be measured.

A rise in the temperature of Th_1 causes its resistance to rise and lower the voltage at one of the amplifier inputs. The amplifier output now changes in proportion to the rise in temperature. The addition of a thermistor to the bias network of an amplifier can be used to stabilize it against the effects of temperature change. Another application is to provide temperature compensation for the change in the winding resistance of alternators and other generators which affects their performance when the operating temperature rises.

Thermo-couples

When two dissimilar metals are in contact with each other a contact potential is developed between them. This is known as Seebeck or thermoelectric effect. The voltage, which rises with temperature, is almost linear over several hundred degrees. If two junctions are formed as shown in Figure 3.13(a), a current will flow around the circuit provided that each junction is at a different temperature. The circuit can be modified as shown in Figure 3.13(b) to include a meter which now becomes one of the junctions, the other becoming the temperature sensor. By reversing the meter connections, temperatures below ambient can be measured. The two metals are chosen to maximize the contact potential for a particular temperature range.

The metals used include:

- iron and copper/nickel alloy,
- copper and copper/nickel alloy,
- nickel/chromium alloy and copper/nickel alloy,
- nickel/chromium alloy and nickel/aluminium/manganese alloy.

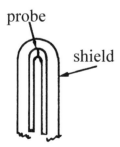

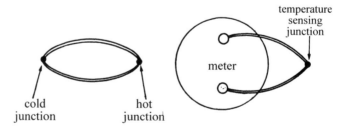

Figure 3.13   Principle of a thermocouple

These allow for a range of instruments to measure temperatures from −85°C up to 2000°C. For most temperature measurements, the hot junction is formed into a probe as shown, left. The metal sheath is used to protect the junction from environmental hazards and is electrically isolated from it with magnesium oxide. This material has good thermal conductivity with high electrical resistance (up to 1000M).

Question 3.7

Thermistors are extensively used as detectors for electronic thermostats to control room heating. Why are thermocouples not used?

Semiconductor temperature sensors

When a silicon diode is forward-biased to operate at a constant current, its junction temperature changes linearly at a rate of $-2mV/°C$. Thus this device can be used as a temperature-to-voltage transducer where the output voltage can be used to indicate temperature.

Magnetic-dependent semiconductors

When a current-carrying conductor is exposed to a magnetic field, a mechanical reaction occurs which tends to cause the conductor to move. This is the principle of the electric motor which was described in the Level-2 book. The direction of the motion can be predicted from Fleming's Left-Hand Rule for motors.

If the conductor is prevented from moving, however, the current carriers flowing within the conductor are deflected to one side. The result is that a voltage is developed across the width of the conductor itself.

In normal conductors this effect is of no practical importance; but when a magnetic field is applied to semiconductor materials, the voltage so generated becomes significant. It is put to practical use in a group of semiconductors called Hall-effect devices, illustrated in schematic form in Figure 3.14 below.

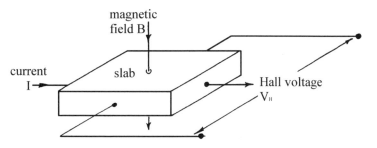

Figure 3.14 Schematic diagram of a Hall-effect device

A Hall-effect device consists of a very small (about 2 mm × 2 mm × 0.5 mm) slab of gallium arsenide, indium antimonide or silicon on to which two pairs of electrodes are evaporated, as shown in Figure 3.14. A control current I and a magnetic field B are applied as shown, and the Hall voltage V_H is developed at right angles to both of them.

The value of V_H is proportional to both the current I and the strength of the magnetic field B, the constant of proportionality depending on the dimensions and nature of the semiconductor material.

Given a constant current, the Hall voltage is directly proportional to the strength of the magnetic field. The first and most obvious application for the device is therefore the measurement of the field strengths of both permanent magnets and electromagnets. It is also used in conjunction with switching transistors in d.c. electric motors, where it can replace the mechanical brush gear which is so prone to wear and to the generation of radio interference by arcing. The device can also be used in contact-less switches, where the movement of a permanent magnet sets up the Hall voltage, which in turn triggers a semiconductor switch.

Magnetic-dependent resistors

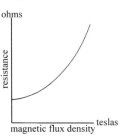

When the application of the magnetic field in the Hall-effect device deflects the current carriers to one side of the conductor, they are caused to flow through a smaller cross-sectional area. This increases the effective resistance of the device, and so reduces current flow. The characteristic of a typical magnetic-dependent resistor is shown, left. Over a normal range of magnetic fields strengths, device resistance can in this way be made to vary by a factor of about 5.

One particularly useful application of a magnetic-dependent resistor is in the 'clamp-on' type of current meter used to measure current flow in power supplies drawing power directly from the electrical mains.

Electro-static discharges (ESD).

NOTE: This topic occurs also in Unit 2, Outcome 2.3 and 2.6 and Unit 4, Outcome 4.4).

Lightning and ionospheric storms are the chief sources of high levels of ESD, but man made electrostatic charges are rather more subtle. These can be generated by the rubbing together of two dissimilar material. Actions such as walking across a carpet or even the nature of the clothing worn, can give rise to *triboelectricity*, as this phenomenon is known.

A triboelectric series lists the materials ranging from air, hands, asbestos, at the most positive, through the metals, to silicon and Teflon as the most negative. The further apart in this table, then the higher will be the electric charge between two materials when they are rubbed together.

MOSFETs, CMOS and Group iii/v (GaAs, etc.) semiconductor devices can easily be damaged by the discharge of static electricity. A discharge from a potential difference as low as 50 volts can cause component degradation, and since the effect is cumulative, repeated small discharges can lead to ultimate failure. It is therefore important that any sensitive equipment should be serviced in a workshop where static electricity can be controlled. Bi-polar transistors are rather more robust in this respect.

The basic work station should provide for operators, work bench, floor mat, test equipment and device under service to be at the same electrical potential. Operators should be connected to the work bench via a wrist strap. To avoid static electricity the operator's clothing also needs to be considered. The wearing of wool and man-made fibres such as nylon creates considerable static. One very useful garment is a smock made of polyester fabric, interwoven with conductive carbon fibres. This has through-the-cuff earthing. The use of compressed air to clean down boards can actually generate static and this can be avoided by the use of an ionized air blast.

Static-sensitive components should be stored in conductive film or trays until required and then handled with short-circuited leads until finally in circuit. Soldering iron bit potentials can also be troublesome unless adequately earthed.

The work bench surface should be clean, hard, durable and capable of dissipating any static charge quickly. These static-free properties should not change with handling, cleaning/rubbing or with ambient humidity.

The use of an ionized air ventilation scheme can be an advantage. Large quantities of negatively and positively charged air molecules can quickly neutralize unwanted static charges over quite a significant area.

Capacitors, including the capacitance of cathode ray tubes should not be discharged too rapidly to avoid damage through excessive current flow. Normal power supply capacitors can be safely discharged through a 1K resistor, but for the EHT voltages found on the CRT, it is safer to use a 100K resistor.

Answers to questions

3.1   400 μA.

3.2   100 mA/V.

3.3   Transistors can be destroyed by ionizing radiation and large electrostatic discharges caused by a nuclear explosion.

3.4   The thyristor or its trigger circuit has failed, allowing the thyristor to operate like a diode.

3.5   The LDR can handle more power, needing no amplification

3.6   The power density is low, so that acres of land would need to be covered by solar cells to produce a useable output. This also makes costs very high.

3.7   The thermocouple detector needs a cold junction and the output is very low.

Unit 2

Outcomes

1. Demonstrate an understanding of analogue and digital meters.
2. Demonstrate an understanding of oscilloscopes and their uses.
3. Demonstrate an understanding of test and measurement instruments for components and signals.
4. Demonstrate an understanding of the role of the PC in test and measurement.
5. Demonstrate an understanding of equipment reliability and surface mount repair.
6. Demonstrate an understanding of safety testing and electromagnetic compatibility.

4

Measurements and readings

Test and measurement

Since the servicing of all electronic circuits calls for the use of measuring instruments, it is necessary to be able to make effective use of these and of the readings obtained from them.

> The two most important types of circuit measurements are d.c. voltage and current readings, using a multi-range meter (multimeter), and signal waveform measurements that are best made using a cathode-ray oscilloscope (CRO, see Chapter 5). One preliminary point of importance is that, whatever the type of measurement made in an electrical or electronic unit, the very act of connecting the instrument into the circuit will have some influence on it and so affect the reading obtained.

Apart from the ease of reading an instrument and its ability to produce consistent readings, the most important features of any test instrument are its accuracy, resolution and linearity.

Accuracy: The accuracy of any meter is quoted in percentage of error relative to some particular standard value. If a meter has a declared accuracy of 5 per cent (say 5 mV in 100 mV), there is little point in trying to interpret a reading to an accuracy of 1 mV. As a useful guide, read only to the nearest half division on the indicated scale. A high grade analogue meter usually has a mirror backed scale plate so that the pointer can be accurately read by avoiding parallax errors. The operator's head is moved from side to side so the reflection just disappears behind the pointer. A digital multimeter often has a better accuracy than an analogue one that employs a moving coil unit. For general purpose instruments, an analogue multimeter might have a declared accuracy of 0.5 per cent, whilst the corresponding digital instrument might produce a figure of 0.025 per cent. However, the least significant digit often varies whilst taking a reading so that the accuracy would typically be declared as 0.x per cent ±2 counts.

Resolution: This term refers to the smallest distinction between readings that can be obtained. For the analogue instrument this is typically the nearest half division mentioned above. For the digital device it is the value represented by the least significant digit.

Linearity: This feature represents a scaling that varies by equal deflections per unit of measured quantity. For an analogue instrument with poor linearity this is usually obvious by the cramping of scale readings towards one end. The linearity of a digital multimeter is not similarly obvious from the reading of the display but can still exist as an unwanted feature of the measurement system.

Periodically, the accuracy of all instruments should be checked against some standard voltage cell or an instrument that has been similarly calibrated and can be used as a *transfer standard*. Since this property is often temperature dependent it is usual to make any readjustment at 20°C. The calibration against a transfer standard meter can easily be affected by connecting the meters to be compared either in parallel (voltage) or in series (current) when coupled to the same source.

Panel meters

Unlike multi-range instruments used for service purposes, panel meters are often built into control systems monitors. Whilst these meters might well be scaled in volts, amps or ohms, they may equally well be scaled in bolts, nuts or any other parameter such as litres of liquid flow per second. They may also contain a limited number of shunts and multiplier so that the single meter can be switched to indicate a small number of different parameters. Otherwise, panel meters are simply basic instruments used for a very specific purpose.

Multimeters

As the name suggests, these are instruments capable of measuring several ranges and parameters. Often d.c. voltage and current, resistance and a.c. voltage. However, only a very few multimeters now include a.c. current scales, because such readings have an acceptable accuracy only if the meter includes an expensive current transformer.

Both analogue and digital multimeters are in general use (see the illustration, left). Analogue multimeters use a moving-coil movement which because of its resistance, draws some current from the circuit under test. Digital multimeters, on the other hand, contain no moving parts; the reading appears as a figure displayed on a readout similar to that of a calculator. A separate power supply, usually a battery, is used, so that practically no current is drawn from the circuit under test. Either type of meter will have a measurable input impedance whose value will affect the readings.

analogue

digital

Activity 4.1

Connect the circuits shown, left, and use first an analogue and then a digital multimeter, each set for a 10 V range, to measure the voltage V. Note the results, and check by calculation that the value of V ought to be 4.5 V when the meter is not connected. Are the meter readings significantly different for (a) the low-resistance circuit, (b) the high-resistance circuit?

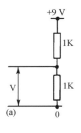
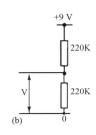

Question 4.1

Given equal accuracy, what is the most pressing reason for using a digital multimeter in a servicing situation?

The effect of meters on circuits

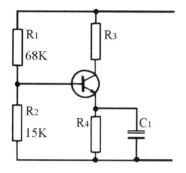

As the results of Activity 4.1 show, the lower resistance of the analogue meter can cause the readings taken to be unreliable. These voltage readings will always be in doubt when the resistance of the meter itself is not high compared with the resistance across which the meter is connected.

The illustration shows a typical bias circuit for a transistor. If the voltage at the base is measured by connecting the multimeter between the base of the transistor and supply negative some current will flow through the resistance of the meter because this is connected in parallel with the resistor R_2.

For reliable readings, the resistance of the multimeter must be high compared with the value of R_2. At least ten times higher is the minimum, and even at this level, you can expect to find some difference between the true voltage (when no meter is present) and the measured value.

Example: A transistor bias circuit having a 9 V supply consists of a 68K and a 15K resistor arranged as in the illustration, left.

(a) What is the bias voltage on the base (assuming negligible base current)?

(b) What voltage will be measured by a voltmeter whose resistance is 150K?

Solution: The circuit is a potential divider.

(a) For such a circuit, $V = ER_2/(R_1 + R_2)$, with $E = 9$ V, $R_1 = 68$K and $R_2 = 15$K, $V = (9 \times 15)/(68 + 15) = 1.63$ V, which is the base bias voltage.

(b) The meter will be connected in parallel with the 15K resistor, so that the combined resistance will be $(15 \times 150)/(15 + 150) = 13.6$K. This quantity must now replace R_2 in the formula above, and the reading becomes:

$V = ER_2/(R_1 + R_2)$ with $E = 9$ V, $R_1 = 68$K and $R_2 = 13.6$K,

$V = (9 \times 13.6)/(68 + 13.6) = 1.5$ V.

Thus the difference is 0.13 V, which is about 8 per cent low.

Range extension using multipliers and shunts

The basic meter resistance R_m is that of the deflection coil which is not only temperature dependent, but can also vary due to manufacturing tolerances. Typically, the resistance of this coil might be 100Ω and this can vary between 90Ω to 110Ω, i.e. a tolerance of ± 10 per cent. If a high stability resistor of 900Ω is wired in series with the meter so the total meter resistance is now 1000Ω, the $\pm 10\Omega$ variation is now reduced to a tolerance of ± 1 per cent. Because of its action this component is referred to as a *swamp* resistor and now forms part of the total meter resistance. The meter shown opposite is assumed to have the following parameters:

Full scale deflection (FSD) current = 50 µA and total meter resistance = 1000Ω.

Using the formula $V = I \times R$, the meter thus has an FSD voltage of 50 mV.

The range of the meter measurement capability can be extended by adding series voltage multiplier (a) or current shunt (b) resistors as shown

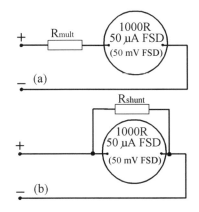

(a)

(b)

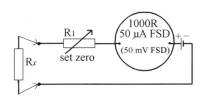

Resistance measurement

in the drawing, left. By dropping an excess voltage across R_{mult} or by shunting excess current away from the meter by R_{shunt}, the meter scaling becomes modified by some multiplying factor n.

For a voltage multiplying factor of n, the meter itself must drop 1 unit of voltage whilst R_{mult} must drop $(n-1)$ units. Thus R_{mult} must have a value of $R_m \times (n-1)$. Similarly for the current shunt, if the meter passes 1 unit of current, the shunt R_{shunt} must pass $(n-1)$ units of current. R_{shunt} must therefore have a resistance of $R_m/(n-1)$. To summarize, $R_{mult} = R_m(n-1)$ and $R_{shunt} = R_m/(n-1)$, where R_m is the total resistance of the meter coil and swamp resistor.

- The two formulae can easily be confirmed by applying Ohm's law to both a series and a parallel circuit.

When the analogue meter is used for voltage readings, it is necessary to know the resistance of the meter for each voltage range. The meter resistance can be found from the 'ohm-per-volt' (Ω/V) figure, or 'figure-of-merit' (also referred to the meter sensitivity), which is printed either on the meter itself or in its instruction booklet. To find the resistance of the meter on a given voltage range, multiply the figure-of-merit by the voltage of the required range.

Example: What is the resistance of a 20 KΩ/V voltmeter on its 10 V range?

Solution: Meter resistance is 20KΩ/V × 10V = 200K.

If it is necessary to take a reading across a high impedance circuit, a high-resistance meter range must be used. This involves employing a higher meter range than the one which seems to be called for. For example, if a voltage of around 9 V is to be measured and the 10 V range of the meter has too low a resistance, the 50 V or even the 100 V range can be used to reduce the distorting effect of the meter on the circuit. Note, however, that if a voltage as low as 1.5V were to be measured, the use of the 100 V range would make this virtually impossible.

The basic analogue meter can be adapted in the manner shown here to add a resistance range to the multimeter circuit. The battery, usually a single cell of 1.5 volts, is used to drive a series current through the meter and the resistance under test. The variable resistor R_1 is used to set the meter to FSD with a short circuit across the test terminals which represents the point of zero ohms (0Ω). The open circuit condition should produce a meter deflection of zero and this represents infinite resistance, $\infty\Omega$. The meter in fact now measures in the reverse direction.

The value of the set zero potentiometer can be calculated as follows:

The volts drop across a meter with an FSD of 50 μA and 1000Ω internal resistance is 50 mV, ($I \times R$). With a 1.5 V cell, there must therefore now be 1.45 volts dropped across R_1. The total circuit resistance R = V/I = 1.5/50μA = 30K. Since the meter resistance is 1000R the value of R_1 must be at least 29K (1.45 volts/50 μA). In practice, a value of at least 30K would be employed.

The range of such an instrument can be modified simply by increasing the battery voltage to, say, 3 volts. Meters of this type normally have a positive potential on the black test lead and vice versa. It is thus important to recognize this point particularly when measuring non-linear devices such as diode and transistors.

Example: Calculate the value of resistance indicated when the meter is at ½ FSD.

Solution: At ½ FSD the meter current is 25 μA so that the total circuit resistance is given by R = 1.5/25 μA = 60K, of which the meter and R_1 account for 30K. There is thus the equivalent of a 30K resistor across the test terminals.

Question 4.2

What types of circuits are most adversely affected by carrying out measurements on working equipment?

Activity 4.2

Calculate the resistance value being measured when the meter above indicates (a) one quarter and (b) three quarters FSD. Comment on the linearity of such an instrument.

Improved accuracy resistance measurements

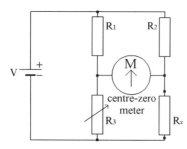

The simple circuit known as the Wheatstone bridge, can provide accurate measurements over a very wide range of values. As shown here it consists simply of 3 resistors, one of which is variable, a centre zero moving coil d.c. meter and terminals across which the unknown resistor is connected. The bridge is energized by a low d.c. voltage (usually a 1.5 V cell). The resistors R_1 and R_2 are referred to as the *ratio arms* which can be switched in or out of circuit as necessary to provide resistance ratios of R_1:R_2 of, typically, 10:1, 100:1 and 1:100, 1:10. The unknown component is connected to provide R_x and the bridge circuit is balanced when there is zero current through the meter (at centre scale).

- The balance condition occurs when $R_1R_x = R_2R_3$, or when $R_x = R_2R_3/R_1$.

The ratio arm values are known and the value of R_3 is found from the scale surrounding the control knob. This allows a simple calculation to be made which provides the value of the unknown to a very high degree of accuracy.

- The bridge type of circuit is used extensively for measurement and is also the basis of the full-wave rectifier circuit.

D.c. meters adapted to read a.c.

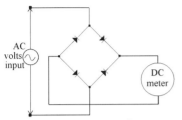

Using a current transformer

By connecting a d.c. moving coil meter to a bridge configured full wave rectifier circuit as shown here, a.c. voltages are converted into d.c. The d.c. meter itself can be fitted with shunt and multipliers so that its basic range can be extended. When used in this way, the meter actually reads the average value of the sine wave input, equal to 63.7 per cent of the peak value. By modifying the scaling of the meter, this can be redrawn to indicate the r.m.s. value instead. Since the r.m.s. value is 70.7 per cent of the peak a.c. value, this scaling or form factor for a sine wave is equal to 0.707/0.637 = 1.11. Different form factors are needed for other waveforms.

When shunts similar to those used for modifying the ranges of a d.c. meter are used for a.c. measurements, their impedance, which varies with changes in the magnitude of the current, would require the use of separate scales for each range. This is the chief reason why a.c. shunts are only found in very low cost instruments. To overcome this problem, a current transformer with a tapped secondary winding is used instead. This has a low impedance primary winding and a high impedance secondary so that the secondary voltage is proportional to the primary current.

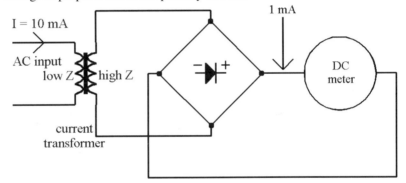

Figure 4.1   A.c. measurements using a current transformer

For example, If the basic 50 µA FSD meter were to be used to measure 5 mA a.c., then the transformer would need to have a 1:100 turns ratio. In this case, the conversion between average and r.m.s. scaling can be accomplished within the current transformer by multiplying the turns ratio by the standard form factor of 1.11. Thus the required turns ratio becomes 1/100 × 1.11 = 1:90. The a.c. output voltage is finally rectified using a standard bridge circuit as shown in Figure 4.1.

Analogue and digital multimeters

Analogue meter features. In theory, the shunts and multipliers used to extend the ranges of basic instruments could be chosen from separate, high stability, high accuracy standard components. However, if these separate resistors are switch selected whilst the meter is left connected to a circuit under test, there is a very good chance that the full test voltage or current can be momentarily applied across the meter with disastrous results. To avoid this situation, shunts and multipliers are generally wound in a series configuration and tapped at the appropriate points. By using such *universal* *shunts or multipliers*, the problem mentioned above can be nearly completely avoided.

Digital meter features. Digital multimeters (DMM) generally have much less effect on the circuit being measured. Most of them have a constant input resistance of 10M or more, and few circuits will be greatly affected by having such a meter connected into them. The operating principle is that the input voltage is applied to a high stability, high-resistance potential divider which is often connected to an operational amplifier (opamp) to provide further isolation. The output of the opamp then provides one of the feeds to a comparator stage (see Figure 4.2). The second comparator input is derived from a stage that generates a sawtooth wave.

At the point in time when the sawtooth starts, a counter also starts and is stopped when the two voltages at the inputs to the comparator become equal. At that moment the count is displayed.

The circuit is arranged so that each digit of the count corresponds to a unit of voltage, say one millivolt, so that the display reading is of voltage. The range switch selects the part of the potential divider to be used, and the position of the decimal point on the display.

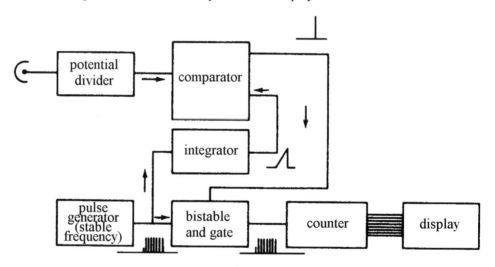

Figure 4.2   Basic block diagram of a digital meter (omitting details of hold-and-measure arrangements)

For example, suppose that the pulse generator or clock of the digital meter runs at 1 kHz, so that 1000 pulses per second are generated. If the integrator is configured so that each pulse produces a voltage rise of 1 mV, then the voltage will rise to 1 V in the time of 1 second. For an input of 0.5 V, 500 pulses will be needed, and the time taken will be 0.5 s. The display will then indicate a reading of 0.500 V. Note, however, that the instrument has taken 0.5 seconds to reach this reading. In general, digital meters are not capable of following rapid changes in voltage. The accuracy of the DMM depends chiefly on the linearity of the sawtooth waveform and the repetition accuracy of the pulse generator.

> Another point to note is that although a digital meter may indicate a voltage reading to several places of decimals, this is not necessarily more precise than the reading on an analogue meter.

DMM devices are often specified by the number of digits in the display. However, there is an extension to this technique to allow for over-range readings. An extra leading digit is either displayed as a 1 or it is not displayed. Such a DMM would be described as having a 3½ digit display. A further addition provides for a minus sign (–) to be displayed when reverse polarity is encountered. This adds an extra ¼ digit. Therefore a basic 3 digit display with these features would be described as a 3¾ digit meter.

Frequency response

This particular property is related to a.c. measurements only. Whilst most multimeters are built to monitor power frequencies, they can provide a reasonably accurate assessment of voltages or current, up to about 20 kHz for an analogue instrument and perhaps 100 kHz for a digital multimeter.

Multimeter facilities

Auto ranging. This feature allows a multimeter to automatically select the correct range of values, when the specific function has been pre-selected by the user. The feature cannot easily be built into an analogue instrument but can readily be included in the facilities of a DMM particularly those instruments that are software controlled. Basically, if the reading on a particular selected scale has been over driven, the meter automatically shifts the scaling up to the next highest level. However, overload can still occur if the meter tries to exceed the maximum permitted level.

Capture of readings. This feature applies chiefly to the DMM and represents the time taken for a reading to reach equilibrium. For the analogue instrument the pointer settles in a time controlled by the damping created by the back e.m.f. due to the pointer movement. In the DMM, this feature can be affected by the dither in the least significant digit (±1 or 2 counts).

Peak readings. As mentioned above, analogue meters actually measure the average value of the input quantity, which in the case of d.c. is also the peak value. However, for a.c. quantities, the scaling is modified by the form or scaling factor. Digital meters can be designed to measure true r.m.s. values for any waveform.

Clamp-on current measurements. When attempting to measure the current flow through such as a large power cable, it is not convenient to break the cable to insert a meter. For these applications there is a device which is effectively a single turn loop that can be opened to encompass the cable and then read off the current flow directly. Such an instrument includes within the clamp a Hall effect device that responds to the magnetic effect of the current flow so that the meter interprets the strength of this field in terms of the current that created it.

Overload protection. There are a number of step that the user can take to avoid damaging a meter through overloading.

- Take note of the manufacturers stated safe working levels for both a.c. and d.c.
- Avoid measuring voltages or currents that exceed the makers recommendations on any particular range.
- Avoid trying to measure volts or current with the meter switched to the resistance range.
- Observe the stated safe environmental working temperature.

Apart from the obvious use of a fuse in the test lead circuit, particularly for current ranges, back-to-back silicon diodes are often included in parallel with the meter movement. These develop a low resistance on overload and thus place a short circuit across the meter movement. Analogue meters may have a ballistic mechanical system that physically trips a pair of contacts to remove it from the circuit if the pointer approaches the end stop too fast. Moving coil meters should always be stored or transported with a short circuit across their terminals. This damps the movement due to the back e.m.f. created by physical movement.

Electronic driven digital or analogue meters can be protected by using the crow-bar effect of a pair of thyristors wired across the input.

Additional facilities. Because of the flexibility of the DMM, its operation can easily be extended to provide measurements of capacitance, frequency, sound level, temperature, humidity and other parameters, as well as allowing for simple diode and transistor testing.

Question 4.3

Why is an a.c. measurement made using a multimeter likely to be inaccurate if the waveform at the measuring point is a sawtooth?

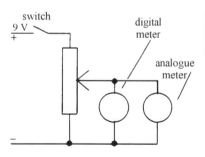

Activity 4.3

Connect two meters, one digital and one analogue, to the centre-tap of the potentiometer, illustrated left. Switch the meters to their 10 V ranges and switch on the 9 V supply. Observe the readings as the potentiometer shaft is rotated to and fro. Which meter most closely follows the changes in the output? Check also how quickly you can read each meter when the voltage is steady. Which meter is easier to read?

Using the multimeter in circuits

The following guide lines could prove valuable.

Start with the meter switched to its highest voltage range.

Connect the meter with the circuits switched off.

For voltage measurements, always use the highest range that gives a readable output.

For current measurements, always try the highest current range first.

Never leave the meter switched to any current range when you have temporarily stopped using it.

Make sure you know which scale on the dial to read before you try to take a reading.

Never leave a meter switched to ohms range when taking voltage or current readings.

Current readings

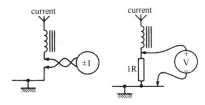

In most circuits, the use of a multimeter for current readings has much less effect on the circuit than when it is used for voltage readings. To make a current reading, however, the circuit has to be broken, and this is seldom easy on modern circuit boards.

A few audio and TV circuits make provision for checking currents by having a low resistance permanently connected to the current path. By measuring the voltage across this resistor, the current can be calculated using Ohm's Law, $I = V/R$. Because the resistance so placed in the circuit is very small, the effect of connecting the meter into the circuit is negligible. The illustration shows these two ways of measuring current.

Voltage readings

Voltage readings are used to check the d.c. conditions in a circuit. These readings are usually made when no signal is present. Readings that are taken when a pulse signal is present will always give misleading results because of the effect of the pulse signal itself on the meter. For this reason, voltage readings shown on a circuit diagram are usually specified as 'no-signal' voltage levels, or are shown only at points in the circuit where the signal is decoupled so that only d.c. is present.

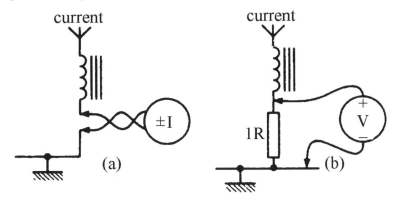

Figure 4.3   Making current readings

In the linear amplifier circuit shown in Figure 4.3(a), the voltage reading at the emitter will not change greatly even if a small signal is present. In Figure 4.3(b), however, emitter voltage will be zero when no signal is present, so that the presence of any signal will cause a voltage of some sort to be recorded. Note that the voltage reading across a point which is bypassed by a large capacitor, as shown in Figure 4.3(a), will always be a purely d.c. reading.

When voltage readings are shown on a circuit diagram, they are always average readings which may vary from one example of a circuit to another because of component tolerances. In addition the actual readings taken in a circuit may be different from these values, either because of tolerances or because of the use of a meter with a different figure-of-merit, even if the circuit is working quite normally. Some experience is needed to decide if a voltage reading which is higher or lower than the stated average value represents a fault, or whether it is simply due to tolerances or meter resistance. In general, a voltage varying from the norm by up to 10 per cent is usually due to tolerances; but a voltage variation that shows a transistor to be nearly bottomed or cut-off in a linear stage always betrays a fault condition.

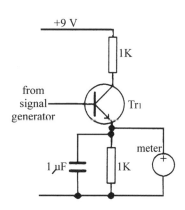

Activity 4.4

Use the circuit shown left, in which the transistor is a general-purpose NPN type such as BC107/108/109, BFY50, 2N3019, etc. Connect the meter, digital or analogue as shown and switch to the 10 V range. Switch on the d.c. supply, and read the meter. Now make readings with a square wave input to the circuit from a signal generator, using 1 kHz square waves starting at 0.5 V p–p amplitude and increasing to 1.5 V p–p amplitude. Observe and note the meter readings for different input amplitudes.

Root mean square (r.m.s.) values

This term has been mentioned previously but without actually defining its important meaning. When d.c. flows through a circuit it dissipates heat due to the power being generated, I^2R watts. Similarly when a.c. flows through a circuit power is again dissipated, but how is this to be compared and measured?

For a sine wave, the average voltage or current flowing will be zero, because each positive excursion is matched by an equal but opposite polarity negative swing. But we know that even a resistor passing a.c. gets hot, so how can we quantify this effect?

The r.m.s. value is described as that value of a sine wave that produces the same power or heating effect as an equivalent d.c. value. For a sine wave, this can be calculated first by squaring the waveform and then taking the average value of this, either of V^2 or I^2. Now by taking the square root of this value either $\sqrt{V^2}$ or $\sqrt{I^2}$ we return to a particular value of V or I which is termed the r.m.s. The process is shown in Figure 4.4.

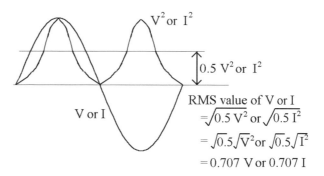

Figure 4.4 The derivation of r.m.s. value for a sine wave

Since the power dissipated or the heating effect is only truly applicable to the sine wave a.c. drive, any other wave shape would produce different results. For example, it can easily be shown using the technique explained above, that the average, peak and r.m.s. values for a square wave are all identical. The effect of all other arbitrary wave shapes can be assessed as follows.

A method that can be adopted for calibrating a non-sinusoidal alternating waveform relies on the basic definition. By using the power of an unknown wave shape to drive an incandescent lamp and a controllable true sinusoid to power an identical lamp, increase the drive to and monitor the sine wave drive r.m.s. level. When the two lamps glow with equal brightness as measured by a light meter or temperature sensor, the r.m.s. value of the arbitrary wave shape is equal to that of the sinusoid drive power.

Non-linearity of scaling

Many of the graphs associated with the non-linear quantities used in electronic engineering have unwieldy long baselines. In addition, these quantities also do not have a linear effect upon the human senses. For example, the typical maximum range of audible frequencies output from an audio system extends from about 20 Hz to 20 kHz or three decades. (Almost 10 octaves, based on doubling of frequency.)

In addition, the audible response effect of the human ear is more nearly related to the doubling of power output levels. For example, if an audio system is set up to produce an output of 1 watt and the ear is allowed time to recognize this level, a doubling of output power would produce the impression of a step increase in output level. Now another doubling to 4 watts would generate the same human audio effect of another step. Continuing this process of doubling power levels would create the impression of a series of linear steps. The audible effect of loudness is therefore also non-linear.

Comparison of audio sound levels is based on a logarithmic scale of Bels (named after Alexander Graham Bell, inventor of the telephone). The Bel (B) which is equal to the logarithm of two power levels, is too large a unit for all normal application, therefore one tenth of Bel or the decibel (dB) is commonly used. As it turns out, this is particularly useful because 1 dB is about the smallest change in power levels that the ear can detect. In calculation terms, the expression is given by:

dB = 10 log (P_1/P_2). where P_1 and P_2 are the two power levels being compared.

If two voltages or currents are both developed in the same value of resistance, then because power is proportional to V^2 or I^2, the above equation can be modified as follows;

dB = 20 log (V_1/V_2) or 20 log (I_1/I_2).

By convention, decibels can take on both positive and negative values which represent either signal level gain or loss in a system performance. The –3dB points on a response curve represent the frequency space between the two half power points and this is usually taken to represent the system bandwidth. The calculation for this value is:

10 log (0.5) = 10 × (–0.301) = –3.01 dB

since 1 dB is about the smallest change in signal level detectable by the ear, this can safely be rounded to –3 dB.

Calculated in a similar way, the half voltage or current level is given by;

20 log (–0.5) = 20 × (–0.301) = –6 dB (approximately)

This relationship can also be explained as follows; power (P) is proportional to V^2 or I^2 therefore 10 log $(0.5V)^2$ = 10 log 0.25 = –0.602 = –6 dB approximately. Thus the half voltage/current points are equivalent to the one quarter power level, –6dB.

To make life easier, it is possible to obtain graph paper with logarithmic/linear (log/lin) and logarithmic/logarithmic (log/log) scales.

Other variants of the decibel are often used in electronic engineering and some of these are defined below:

dBW   decibels relative to a 1 watt level,

dBm   decibels relative to a 1 milliwatt level,

dBV   decibels relative to a 1 volt level,

dBμ   decibels relative to a 1 μvolt level,

dBc   decibels relative to a carrier level,

dBA and dBC variants are used in the comparison of environmental noise power levels.

Activity 4.5

A certain resistance measuring meter has a scale plate calibrated between 0 and 1.0 in 0.1 steps. The following table of results were obtained during calibration.

Meter reading	0.2	0.3	0.4	0.5	0.6	0.7	0.8
R Value (Ω)	2560	1280	640	320	160	80	40

Sketch the graph using linear scales and estimate the ohmic values for the scale readings of 0.25 and 0.55. (Answer approx. 1850R and 250R).

Note how difficult it is to read values between plot points (interpolation) and how it is it is even more difficult to estimate values for readings of zero to 0.1 and 0.9 to FSD by extrapolation (outside the range of plot points).

Activity 4.6

The results tables, below, were obtained by a frequency response experiment on an audio system driver stage.

Plotting the frequency and the output power in octaves and dBm respectively results in a log/log graph. Whilst it is difficult to interpolate values of frequency, the graph is restricted to a more manageable size. The −3dBm points occur at about 50 Hz and 16 kHz so that the bandwidth is approximately 15.950 kHz.

Frequency Hz	20	40	80	160	320	640	1280	2560	5120	10240	20480
Output mW	1.0	5	10	12	13	13	13	13	12.5	10	6
Output dBm	0.0	17	10	10.8	11	11	11	11	10.9	10	7.8

Networked instruments

The concept of automated test equipment (ATE) has been progressively introduced into the servicing environment and this involves simple local area networks (LAN). Not only can the networked system now provide service information to many test and repair stations, such systems can also be used to monitor and log data from various test points for future analysis. Furthermore, the computer monitor can be itself be converted into a range of test instruments all under software control (see also Chapter 7, *Virtual instruments*).

RS232

This is probably the oldest interconnection standard for linking a computer to an item of ancillary equipment. During its more than 25 years existence, the standard has undergone more than six major changes and is still valid today.

The original RS232 standard was designed to provide 2 transmit/receive channels for simultaneous full duplex communications using bipolar signals of ±15 volt level. The positive and negative excursions represented logic 1 and 0 respectively. Using a multi-core cable and 25 pin D-type connectors, a maximum signalling rate of 20 kbits/s could be achieved over a distance of 20 metres. Later versions of RS232 have used both 15 pin and 9 pin D-type connectors with multi-core cable and bipolar signals at both ±3 and ±5 volts levels in a single simplex transmit/receive mode.

In all the RS232 versions, the system transmits in the bit serial mode in both directions and the system standard provides for data, timing and control signals.

It is designed to connect a data terminal equipment (DTE) usually a computer, to a data communication equipment (DCE) or peripheral device. The primary channel is normally used for data, whilst the secondary path carries the control and timing signals. However, both channels may be configured for both half duplex and full duplex transmission modes.

The illustration shows the standard pinning arrangement for the RS232 female socket.

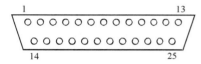

Important signal and control lines

DCD   data carrier detect line, carrier present when logic 1.

RXD   Receive data, serial data input.

TXD   transmission of data, serial data output.

DTR   data terminal ready, a logic 1 indicates that peripheral is ready to connect to the communication channel.

Signal earth   common ground return for both ends of transmission link.

DSR   data set ready, logic "1" indicates that the handshake process is complete.

RTS   request to send output, peripheral transmits data when logic "1" is present.

CTS   clear to send, when logic "1" the peripheral is ready to receive data.

RI   Bell, ring indicator, indicates that "bell" has been detected. A hang over from teleprinter systems.

Table 4.1 Full implementation pin allocation. Few applications now make use of all of these pins

Pin	Function	Pin	Function
1	Power ground	14	Secondary TXD
2	TXD	15	Transmit clock (DCE)
3	RXD	16	Secondary RXD
4	RTS	17	Receive clock
5	CTS	18	Unassigned
6	DSR	19	Secondary RTS
7	Signal ground	20	DTR
8	Data carrier detect (DCD)	21	Signal detect
9	Data set test	22	Bell detect
10	Data set testing	23	Bit rate select
11	Unassigned	24	Transmit clock(DTE)
12	Secondary carrier detect	25	Unassigned
13	Secondary CTS		

Table 4.2   Implementation of 9-pin RS232 version

Pin	Function	Pin	Function
1	DCD	6	DSR
2	RXD	7	RTS
3	TXD	8	CTS
4	DTR	9	Bell
5	Signal earth		

For a modem cable using the full 25 pin implementation, the 2 connectors are configured pin for pin, but for the null or non-modem operation, pins 2 and 3 (RXD and TXD) and pins 5 and 20 (DTR and CTS) are cross connected. The changes for the 9 pin cable are pin to pin for a modem cable and for the null-modem, pins 2 and 3 (RXD and TXD) and pins 4 and 8 (DTR and CTS) are cross connected.

IEEE488 Bus

This is also known as general purpose interface bus (GPIB) and Hewlett Packard instrumentation bus (HPIB) and is a standard interface bus designed specifically for instrumentation purposes. It consists of eight data lines, three control lines and five lines for interface control. It is based on a standard male/female back-to-back 24-pin connector and cable system, that provides for 8-bit parallel data transfer rates of 1 Mbyte/s. The transmission system uses unipolar negative logic signalling (High = 0, Low = 1). Fifteen devices with different data rates may be connected to the bus at any one time, but the total transmission line length is restricted to twenty metres maximum.

Question 4.4

In the RS-232 system, what do the initials DTR mean?

In the latest version, IEEE488.2, the above parameters have been somewhat upgraded. Three types of device may be connected and these are described as either, *talkers, listeners or controllers*. For example, a meter, keyboard or sensor may be described as a talker, a printer or a recorder as a listener and a computer as a talker-listener. By comparison, a DMM that needs to be programmed in situ and then be able to transmit results, is also a talker-listener. Although the network may contain more than one computer, one is designated as a master or controller and allocates the bus to individual devices in turn.

Each device on the bus has its own unique address carried in a specific byte. The five least significant bits (LSB) represent the address, bits 6 and 7 set the talk/listen function, with 01 = listen and 10 = talk. The MSB can generally be ignored.

Table 4.3   IEEE488 pin configuration.

Pin	Signal	Pin	Signal
1	Data I/O 1	13	Data I/O 5
2	Data I/O 2	14	Data I/O 6
3	Data I/O 3	15	Data I/O 7
4	Data I/O 4	16	Data I/O 8
5	EOI	17	REN
6	DAV	18	DAV ground
7	NRFD	19	NRFD ground
8	NDAC	20	NDAC ground
9	IFC	21	IFC ground
10	SRQ	22	SRQ ground
11	ATN	23	ATN ground
12	Shield	24	Logic ground

Key to bus control

ATN (attention) is the main controller signal. When this is low, the controller is sending commands or addresses to the interface connected devices. When it is set high, it signifies to the talkers and listeners that the bus is ready to transfer data.

IFC (interface clear). By driving this signal low for a short period (100 µs) the controller resets the interfaces of all the devices on the bus.

REN (remote enable). This line is held permanently low when the interface is working.

SRQ (service request). Devices use this line to request service from the controller. For example, an instrument can signal that it is ready to read after taking a measurement.

EOI (end or identify). This line is driven low by an active talker whilst it is transmitting its last byte as a form of end of data string. It can also be used by the controller.

DAV (Data valid).

NDAC (Not data accepted).

NFRD (Not ready for data).

Control protocol

When ATN is low (1), all devices must listen for interface messages, then if a device is addressed, it must be ready for a data transfer as soon as ATN goes high (0).

Every byte is transmitted under the control of three handshake lines, DAV, NRFD and NDAC.

An active talker with data to transmit monitors the NRFD line and if this is low (1), the listeners are not ready to receive data. When NFRD goes high (0) and the listeners maintain NDAC low (1), the data is not being accepted. When the listeners set NDAC high (0) and NFRD low (1), the data can be transferred.

At the end of each transmitted byte DAV goes low (1) to validate the data transfer.

Answers to questions

4.1   There are no reading errors caused by parallax.

4.2   High-resistance and RF circuits.

4.3   The meter measures average values and the scaling assumes the use of a sine wave.

4.4   Data terminal ready.

Activity 4.2   (a) 90kΩ and (b) 10kΩ. These results show the most serious problem with this form of resistance measurement.

5   Oscilloscopes

The cathode ray oscilloscope (CRO)

The CRO is a device that is capable of writing a 2 dimensional (2-D and in some cases 3-D) graph on a specially produced glass screen. It does this under the influence of two (or 3) input signals whose effect is to produce a recognizable pattern that describes certain features of the various signals.

The cathode-ray tube (CRT) operates by harnessing the movement of electrons in a vacuum, based on the following principles:

1. Electrons are released into a vacuum when a cathode material, such as barium oxide, is heated to 1000°K or more

2. Electrons, being negatively charged, are attracted to positively charged metal plates or are repelled by negatively charged plates.

3. Moving electrons constitute an electric current, and so can themselves form part of a circuit

4. The direction in which a stream of electrons moves can be changed by applying to it one or more electrostatic or electromagnetic fields.

5. Moving electrons carry energy, and so will cause substances known as phosphors to glow brightly when they are struck by an electron beam.

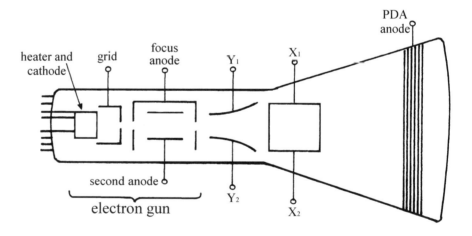

Figure 5.1   CRT for instrumentation applications

A cross-sectional diagram of a typical instrumentation CRT is shown in Figure 5.1 and the function of the various parts is as follows.

The *cathode* is formed as a nickel cup which is coated at its closed end with a mixture of oxides of barium, strontium and calcium. These elements emit electrons when at red heat. The heating function is carried out electrically, using a coiled molybdenum or tungsten *heater* wire which has to be coated with aluminium oxide to insulate the heater from the cathode. Failure of this insulation is a fairly common CRT fault. In a CRO, the heater is commonly driven from a 6.3 volt a.c. supply.

Surrounding the cathode is another, larger, metal cup called the *grid*, or control grid. This has a small (0.025 mm or less) hole at its centre. The heater, cathode and grid constitute the source of the electron beam and require four connections, two for the heater, and one each for the cathode and grid.

The voltage between the cathode and the grid controls the flow of electrons through the hole in the grid. In the cut-off condition, typically at a grid voltage of about 85 V negative with respect to the cathode, no electrons pass through the hole in the grid. As the grid voltage is increased positively to about –5 V relative to the cathode the electron beam current increases progressively to maximum.

> Operation of a CRT with the grid positive to the cathode is normally undesirable, but it can help an elderly tube to sustain electron emission, and is a technique sometimes used in conjunction with higher-than-normal heater currents, in a reactivation process.

The electron stream emerging from the hole in the grid is a diverging stream which then needs to be converged, or focused, so that a spot of light is produced at the exact point where it hits the screen. Focus is achieved by placing, at appropriate places along the length of the gun, metal cylinders (anodes) carrying different voltages. This system is called *electrostatic focusing* and employs variable voltage levels to obtain the desired degree of focus. An alternative technique called *electromagnetic focusing* has been used in the past when a focus coil through which d.c. flowed, was positioned along the neck of the tube.

Once focused, the beam is then deflected so that it can be directed at any part of the screen inner surface. Electrostatic CRTs used in oscilloscopes use metal deflector plates to achieve this. The pair of plates situated closer to the cathode are called the Y-plates, and are used for vertical deflection of the beam. The other pair, mounted at 90° to the plane of the Y-plates, are called the X-plates, and cause horizontal deflection of the beam. Because the Y-plates are nearer to the beam source, the deflection sensitivity, measured as the number of millimetres of spot deflection per volt difference between the plates, is greater for the Y-plates than for the X-plates.

Finally, after focusing and deflection, the electron beam strikes the phosphor material deposited on the inner screen face. This material, which is an insulator, can be coated with a thin film of aluminium which acts in two ways:

> It provides a metal contact so that a high (positive) accelerating voltage can be applied to it.
>
> It reflects light from the phosphor which would otherwise be lost into the tube.

Electrons can penetrate this aluminium layer quite easily, so that the gains to be had from using such a layer easily exceed the losses.

The voltage which is applied between the cathode and the screen (the final anode connection) is called the EHT (which stands for extra high tension). It has the effect of accelerating the electrons towards the screen. The greater this acceleration, the brighter will be the spot when the full beam current strikes a phosphor dot on the screen. The value of the accelerating voltage also has an effect on the deflection sensitivity. At large values of EHT, much more deflection effort in terms of voltage between the plates, is needed to achieve the same number of millimetres of deflection than is needed at lower EHT values.

Electrostatic deflection is well suited to signals which can range in frequency from d.c. to a several hundred MHz, because the deflection plates behave in a circuit as a high impedance with a small-value capacitance. Since no d.c. current is drawn by the plates, they can be driven by signals from voltage amplifiers.

In order to maintain the bright traces that are needed for very fast deflection systems, a technique known as post deflection acceleration (PDA) can be employed. This can be achieved by printing a coil of graphite or nickel alloy around the end of the tube close to the screen in the manner shown in Figure 5.1. A CRT using this technique operates with a second/third anode voltage of around 1000 V and a PDA anode voltage at least twice this value and even up to 15 kV.

Typical operating voltages

Table 5.1 below compares a typical range of voltages encountered on the CRT electrodes used with a CRO and a typical monochrome television tube, (in practice, values may differ significantly from those quoted here). In order to allow the grid to operate near to earth potential, the CRO conveniently uses a combination of negative and positive supplies to obtain the high potential required between cathode and final anode.

Table 5.1 Typical operating voltages

Type	Cathode	Grid	1st anode	Focus anode	Final anode
Oscilloscope	−100 V	0 V	+100 V	0 to −200 V	2 to 4 kV
Monochrome TV	+80 V	+20 V	+300 V	up to 350 V	10 to 15 kV

To overcome the transient nature of the display, the CRO can be fitted with a camera attachment to provide a permanent record. Each oscilloscope type is commonly available with a range of alternate CRTs, each with a different persistence of illumination and colour. Table 5.2 provides a useful guide to their particular application:

Table 5.2 Monochrome CRT phosphor characteristics

Persistence	Colour	General application
Long persistence	Orange	Low-frequency displays
Medium persistence	Green	General-purpose displays
Short persistence	Blue	High-frequency displays

Question 5.1

Why should a cathode ray oscilloscope be operated clear of magnets or inductors connected to a.c. supplies?

Common CRT faults

The following are some typical faults.

1. Low maximum brightness, the causes could be either low EHT, faulty voltage supply to the grid or low cathode emission.

2. Uncontrollable brightness, caused by shorts between either the heater and cathode or grid and cathode.

3. Deflection irregularities, usually caused by amplifier failures, but if an instrument has been roughly handled, a deflection plate could have become displaced.

4. No trace, probably caused by amplifier or power supply failure, or by an o/c tube heater.

5. Other tube faults, include misaligned grid, o/c connections, and leakage of air into the tube. A leaking or soft tube (down-to-air) can be detected if the gettering (the name used to describe the normally silvery deposit at the neck of the tube) has turned white.

Activity 5.1

Examine an assortment of electron guns from instrumentation CRTs, both single and dual trace, identify and note the position and structure of the various electrodes.

At the heart of the display system is the sawtooth waveform that is used to produce the horizontal beam deflection across the tube face. This causes the beam to traverse the screen relatively slowly during the forward writing period and then rapidly fly back to repeat the process in a continuous manner. Whilst this is in progress, the signal to be examined, the work signal, is applied to the vertical deflection system so that a 2-D pattern that describes its amplitude, periodic time and general shape can be displayed. Typical basic beam-deflection sensitivities, which are accurately known for each tube, vary between 0.02 and 0.05 cm/V.

Activity 5.2

Use a single beam oscilloscope with no input signal. With all supplies to the tube switched off and all leads shorted one after another to earth in order to discharge any capacitors, attach the leads of a high resistance voltmeter to the cathode (+) and grid (−). Replace the covers so that no points where high voltages will be present can be touched when the tube is switched on. **It should be made impossible for you or anyone else to touch any part of the voltmeter leads save where they are properly insulated**.

Switch on, and note the action of the brightness control. If a photographic exposure meter is available, plot a graph of screen brightness (representative of beam current) against bias voltage.

Multi-trace CROs

A CRO with a single-beam CRT can function as a multi-trace CRO in one of two modes, chopped or alternate. In the chopped mode, the beam is rapidly switched many times per trace cycle between two channels to provide a double-trace display. Switching high-frequency signals in this manner produces unacceptable gaps in the displayed waveform and thus this mode is limited to use at relatively low frequencies. In the alternate mode, one sweep is used alternately for each channel, but at the expense of reduced trace brightness.

True double-beam CRTs employ tubes with two identical electron guns mounted side-by-side within a common glass envelope. These can also be adapted as above to provide four channel operation.

An earlier type of double beam CRT added an extra deflection plate mounted in parallel with and between the two Y-plates. This was held at ground or a slightly negative potential so that as the electron beam approached this region, it became split into two paths which could be further deflected independently by signals applied to the separate Y-plates. In this system, pioneered by Cossor in the 1940s, one trace is inverted with respect to the other.

Question 5.2

What is the name of the technique used on a CRT to maintain high trace brightness without decreasing deflection sensitivity?

The analogue oscilloscope

The signal to be examined is input to the Y amplifier section via a calibrated step attenuator which commonly provides an input impedance ranging from 1 to 10M in parallel with a self capacitance of 10 to 20 pF. This stage typically has a rise time of a few tens of nanoseconds. (Rise time is the time taken for the signal to rise or fall between its 10 per cent and 90 per cent levels.) Thus the input bandwidth can easily exceed 100 MHz. Dual channel CROs will be equipped with two such stages. The major part of the Y channel gain is achieved in the driver and output stages to the Y-plates.

The X deflection system is driven from a sawtooth generator stage that provides a waveform with very low distortion. This signal is further amplified before being used to drive the X deflection system. Again, for a dual or double beam instrument there will be two similar display driver stages, but generally driven from one (or possibly two) timebase circuits. The timebase sawtooth can be synchronized to either of the two input work signals (Y) selected by a switch as shown in Figure 5.2. Provision is also commonly available to compare the Y input with an external signal and therefore an additional switched input can be provided as shown.

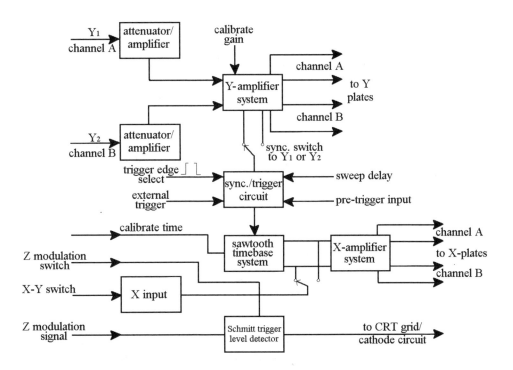

Figure 5.2   Block diagram of two-channel analogue oscilloscope

D.c. shift potentials are also provided for both the X and Y channels to position the trace precisely within the graticule scale that is commonly ruled in 1 cm squares.

At times, it is necessary to study just a short time scale section of the Y signal and this can be done by blanking this input until such times as the important section becomes available. This can then be studied by illuminating the trace through a bright-up signal known as the *Z modulation* so that only the wanted time period is displayed. The effect is almost that of a 3-D display.

The visual display is supported by a scribed and sometimes illuminated graticule. This allows fairly accurate assessments of amplitude, phase relationship and time to be made. In addition, with experience it is possible to obtain a reasonable assessment of any distortion found in a waveform.

Functions

Normally the Y input is coupled via a high value of capacitor to ensure that the frequency response can extend as low as 10 Hz. The CRO is equipped with a switch labelled **a.c. / d.c. / ground** and in the **d.c.** position the coupling capacitor is short circuited. This is provided so that the Y input signal that has a d.c. component can be displayed correctly. The **ground** position is useful because it sets the unmodulated horizontal trace at ground voltage level which allows this axis to be accurately positioned on a particular graticule line referenced as zero volts.

Even with relatively low cost oscilloscopes several variations of the basic synchronism (sync.) system can be provided. The timebase can be synchronized from an external source, from either the rising or falling edges of the Y input, or even a range of input signal levels in order to avoid false sync through noise or glitches. When a second auxiliary timebase is provided, this can be used as a delay to sweep the sync system through sections of very long waveforms in order to examine short period intervals of interest. Again, in some oscilloscopes *pre-triggering* is provided so that it is possible to examine part of a waveform that would appear before the normal sync pulse.

Like all test instruments, it is important to be able to rely on the accuracy of the readings obtained. The CRO usually has a built in calibrator that periodically needs to be checked against some standard. The sub-standard typically consists of an internally generated square wave with a frequency of 1 kHz and amplitude of 1 volt. When used as the work input, the gain of the Y amplifier can be preset to indicate an amplitude of 1 volt. The periodic time for this waveform is 1 millisecond and thus the duration of one timebase scan can be similarly adjusted.

Input impedance

The resistive component of the input impedance will load the d.c. conditions of a circuit, whilst the capacitive component can distort the signal waveform to be examined. Whilst the total input impedance at the Y sockets can be around 10M in parallel with 20 pF, this can be seriously compromised by the cable used to connect to the circuit under test. The use of a low loss coaxial cable that is too long can easily distort the leading and trailing edges of square waves. Even the sine-wave frequency response can be affected because the combination of circuit output resistance and CRO input capacitance can act as an integrator or low-pass filter.

When the CRO is used to measure pulse waveforms in medium to high-resistance circuits, a low-capacitance probe should be used. Such probes are available as extras for most types of oscilloscopes. A few probes are active in that they contain transistor or FET amplifiers, but most are passive, containing only a variable capacitor and a resistor as shown, (a), left. The input voltage is divided down, so that a more sensitive voltage range must be selected; but the effect of capacitance is greatly reduced because the capacitance of both the cable and the CRO input is used as part of the divider chain.

In the equivalent circuit of the low-capacitance probe (b), C_2 includes the stray capacitances of the cable and of the oscilloscope. C_1 is varied until $R_1C_1 = R_2C_2$. The signal attenuation which is given by the expression:

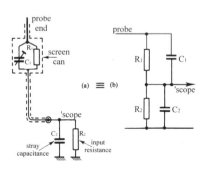

$R_2/(R_1 + R_2)$ shows how it is necessary to select a more sensitive oscilloscope range setting. The probe is normally calibrated against the inbuilt 1 kHz 1 volt peak-to-peak internal square wave calibration source.

The basic probe arrangement illustrated here can be extended to provide a ×1/×10 input attenuator probe by including a switched resistor network. Again this device has to be calibrated in a similar manner to that described above.

Measurements using the CRO
Voltage

To measure the peak-to-peak amplitude of a signal, the vertical distance, in cm, between the positive and negative peaks must be taken, using the graticule divisions. This distance (y in Figure 5.3), is then multiplied by the figure of sensitivity set on the **Volts/cm** input sensitivity control.

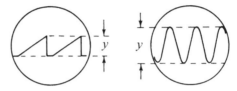

Figure 5.3   Measuring peak-to-peak amplitude

Frequency and periodic time

To measure the duration of a cycle of an a.c. signal, its periodic time, between successive positive or negative peaks, the horizontal distance between them is taken, using the graticule scale. This distance in cm is then multiplied by the time value read off the **Time/cm** switch scale. The frequency of the wave can be calculated from the formula:

frequency = 1/(periodic time).

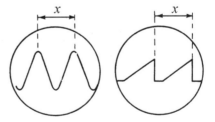

Figure 5.4   Measuring the periodic time of waveforms

With time measured in units of seconds, the calculated frequency will be given by the formula in Hertz (Hz). If the time is measured in milliseconds (ms), the frequency will be in units of kilohertz (kHz); if the time is in microseconds (μs), the frequency will be in Megahertz (MHz).

In some older oscilloscopes, a continuously variable **Time/cm** control needs to be turned to one end of its travel when time measurements are made, and any **X-gain** control has to be at its minimum setting.

Current

Low impedance electromagnetic scan coils have in the past been used to monitor current flow, but will only be found in very specialized oscilloscopes. In this case, the coils were externally mounted over the main electrostatic deflection system. For the measurement of current it is much more common to monitor the voltage developed across a known value of resistance. The current is then evaluated by applying Ohm's law.

Activity 5.3

In order to utilize the CRO to maximum advantage, especially its more sophisticated modes of operation, it is important to obtain as much real time practice as possible.

Connect a signal-generator to the input terminals of the CRO. Set the signal generator so as to provide a 1 kHz square wave of 1 V p–p. Adjust the CRO so as to obtain a locked waveform, and read the amplitude and time. Does the time reading correspond to the frequency as set on the signal generator? Compare these settings with those obtained from the inbuilt calibration signal.

Change the signal generator waveform, amplitude and frequency settings, and measure the new waveform's amplitude and periodic time on the CRO. Check that these values agree with the generator settings.

External triggering

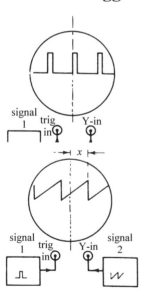

An oscilloscope fitted with an external trigger (EXT. TRIG.) input can be used for comparing the phases of two waves if a double-beam CRO is not available. Feed a signal into the EXT. TRIG. input, with the trigger selector switch set to EXT, this will cause the timebase to be triggered by that waveform. The timebase will now always start at the same point in the waveform. The triggering wave can be seen on the screen by connecting the Y-INPUT socket to the same source. The X and Y shifts can be used to locate one peak of the wave over the centre of the graticule, see the illustration, left which shows one leading edge of a pulse coinciding with the vertical line of the graticule.

Now disconnect the Y-INPUT from the first waveform and replace it by the second source at the same frequency. A locked trace will appear on the screen, as illustrated. If there is a time difference between the two waveforms, the peak of the second wave will not be over the centre of the screen, because the timebase is still being triggered by the first waveform.

Measuring the distance X horizontally from the centre allows the calculation of the time shift either earlier (left of centre) or later (right of centre) of the second waveform as compared with the first. This time difference can be converted into phase angle if the two waveforms are sine waves.

The conversion formula is: $\theta = (360 \times t)/T$,

where θ is the phase angle in degrees, t is the time difference and T is the time of a complete cycle expressed in the same units as t.

Example: A complete cycle of a waveform takes 3 ms, and a second wave has its peak shifted by 0.5 ms. What is the phase difference between the two waves?

Solution:

Using the formula above $\theta = (360 \times 0.5)/3 = 600$.

Phase differences using Lissajou's figures

Phase differences between two sine waves of the same frequency can also be measured using the well-known technique of Lissajou's figures, which represents an application of the use of a non-linear timebase.

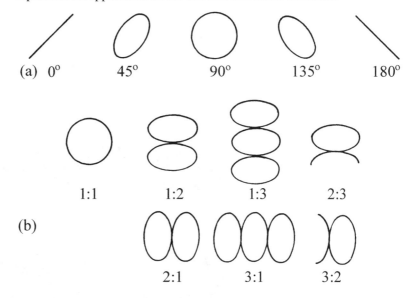

(a) $0°$ $45°$ $90°$ $135°$ $180°$

(b)

1:1   1:2   1:3   2:3

2:1   3:1   3:2

Figure 5.5 Lissajou's figures (a) varying phase shift for $f_y/f_x = 1$, (b) varying ratios of $fy:fx$ for a $90°$ phase shift

A series of such displays is shown in Figure 5.5. If two sine waves are input to X and Y deflection systems and their amplitudes are carefully adjusted, the display will vary from a straight line, through a series of ellipses, to a circle in the manner indicated. Furthermore, this technique leads to the more complex displays shown in Figure 5.5(b), when the two waveforms are simple multiples of each other. In general, the ratio of the two frequencies here is given by:

f_y/f_x = (Number of loops horizontally/Number of loops vertically)

The Lissajou's figure technique can be extended to determine an unknown frequency. This is achieved by making a direct comparison of the unknown with the output from an accurately calibrated sine-wave signal generator.

Activity 5.4

Periodically, both the X and Y amplifier circuits need to be re-calibrated. Carry out this operation using the internal square wave standard as follows.

Check the bandwidth of the Y amplifier chain by proceeding as follows. Using a standard sine wave signal generator, provide an input signal with an amplitude to produce a 5 cm peak-to-peak trace. Progressively increase the generator output frequency whilst maintaining the same input level and note the frequency at which the amplitude falls to $5 \times 0.707 = 3.535$ cm (approx. 3.5 cm). This is the upper cut-off frequency, and since the Y amplifier must have a d.c. response, it is also the amplifier bandwidth. Check that this figure agrees with that stated in the instrument service manual.

Question 5.3

What connections are used to the CRO to obtain a Lissajou figure?

Activity 5.5

Check the accuracy of the amplitude of the internal standard waveform as follows. This can be checked against a standard Weston cell which provides a d.c. e.m.f. of 1.0186 volts at 20°C. When connected to the Y input this will produce a deflection of the horizontal trace by marginally more than 1 V/cm at about 1% accuracy. If a standard cell is not available, then a new 1.5 V mercury dry cell can be used. This will produce 1.5 cm deflection at the same setting with a reasonable degree of accuracy.

Digital measurements

Processing a digital signal will create pulse distortion. What starts as a train of square pulses rapidly approaches sinusoids as the high-frequency components are progressively removed. Again, Lissajou's figures can be used to quantify this distortion. The basic principle is shown in Figure 5.6(a), where a digital data stream is applied to the CRO Y input. A sinusoidal timebase is provided at the external X input, running at a quarter of the bit rate.

When the phasing of the two inputs is correctly adjusted, an *eye pattern* appears as indicated in Figure 5.6(b). The upper and lower levels of this trace, which represents the levels produced by a run of several consecutive

1s and 0s in the data stream, is taken as the maximum eye opening. A large amplitude of eye opening represents a low level of data distortion and hence a low bit error rate (BER). The ratio of (a/b) × 100% can therefore be used to evaluate the quality of the data signal at various points in the signal-processing chain.

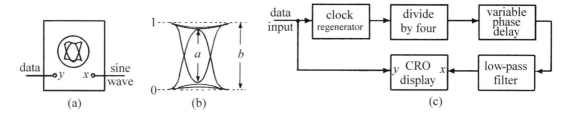

(a) (b) (c)

Figure 5.6 Producing eye diagrams

Figure 5.6(c) shows the classical way in which an eye height display can be produced. The original data stream is used first to lock a local clock oscillator circuit to the bit rate. This is then divided by four and low-pass filtered to provide the one-quarter rate timebase frequency. The data and sinusoid are then input to the CRO as described above. An eye pattern will be displayed when the variable phase delay is correctly adjusted.

Storage oscilloscopes

Storage oscilloscopes are used to display fast one-shot, transient events or very low frequencies that change too slowly to be captured by the normal-persistence CRO tubes. These instruments are available in either digital or analogue form.

In the digital form, the signal of interest is first captured, then digitized and stored in a semiconductor memory. From here it is readily available for future display and analysis.

The analogue storage tube carries a fairly conventional electron gun structure, but this writes an electron pattern on a layer of insulator deposited on a mesh screen placed close to the tube face. Once it is captured, a second gun (the flood-gun), unaffected by the deflection system, can be energized to flood the whole of the mesh area with electrons.

This flood gun uses low-velocity electrons, and where the insulator on the mesh carried a larger negative charge, the flood gun beam is repelled and does not affect the screen. In other words, the mesh acts as a form of large grid that controls the flow of flood gun electrons to the screen. This causes the pattern of charge on the mesh to be displayed as a corresponding pattern of brightness on the phosphor at the tube face, and this will persist until the mesh charge is cleared, so erasing the trace.

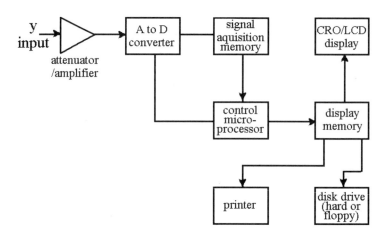

Figure 5.7   Block diagram of a digital storage oscilloscope (DSO)

The digital storage oscilloscope (DSO)

Figure 5.7 shows the basic principles of the operation of a digital storage oscilloscope. The work signal or Y input stages are very similar in circuitry and function to those found in the conventional CRO. High input impedance, high bandwidth and fast rise time are typical of those parameters found in the best of analogue instruments. This input stage carries the familiar calibrated attenuator, together with buffer amplifiers to avoid loading the driving circuits under investigation. This stage is followed by analogue to digital signal conversion before the data is stored in a semiconductor signal acquisition memory, all under the control of the software stored in the control microprocessor. The processed data is then passed through the microprocessor bus system to the display memory where it is organised into a high speed raster scan format for readout and then display on a CRT or LCD screen.

The display data can be stored in a disk memory (either hard or floppy drive) for future replay and investigation. It is also possible to obtain a hard copy print out via a standard printer port.

The digital phosphor storage oscilloscope (DPO)

In this recent development for oscilloscope design, the phosphors are used directly to add a third dimension (3-D) to the display. This is achieved by introducing a variable brightness amplitude control, either in colour or monochrome, unlike the on-off display that is provided through the Z-modulation of an analogue instrument.

As with the digital storage oscilloscope, the front end stages again consist of an amplifier/attenuator and analogue to digital conversion. The next stage stores the data in an intelligent 3-D database called the *digital phosphor*. The data is then passed to the display memory where it is organised into a raster format for periodic transfer to the display device. The screen display not only carries waveform data in pictorial analogue fashion, but also other pertinent information regarding the statistical analysis of the signals being processed. In parallel with these operations, the microprocessor automatically performs measurements and the mathematical functions mentioned above. The block diagram of Figure 5.8 illustrates the principles.

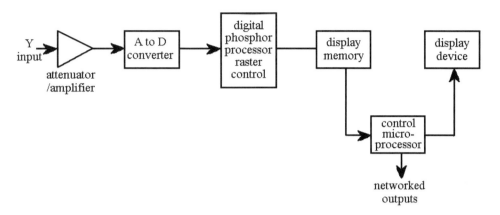

Figure 5.8 Basic block diagram of the Tektronix Inc. digital phosphor oscilloscope (DPO)

Apart from its use in signal processing, the microprocessor also controls the data transfers to the outside world via ports that are Ethernet, IEEE-488 (GPIB), RS232-C, and Centronics compatible.

Analogue and digital comparison

Both the analogue and digital CROs still have their place within the servicing environment because each have advantages that are not found in the other type. The analogue real time (ART) type of instruments have been in use for a long time and their operation and use are thus very well understood. Because the display is in real time, any fleeting signals become very difficult to capture by any recording method, other than a camera. This in itself would create film processing problems if the Polaroid Land camera had not been developed for this purpose. Even with this it is still very difficult to synchronize and photograph a very brief fleeting signal feature unless it is in some way repetitive. The DSO can display and store slow waveforms and short duration transients with equal ease. It also has the ability to store the waveform data and re-run it on demand. The DSO can perform watch-dog facilities such as searching for a glitch or other fleeting signal feature over a long period of time, plus providing multi-channel operation, with automated measurement facilities. The ART provides waveform information that is truly real time, whilst the DSO outputs are delayed somewhat due to signal processing.

The input signal must be sampled at a rate at least twice that of its highest frequency component to avoid aliasing which creates distortion and introduces additional frequencies into the waveform. The quantization rate must be at least 8 bits and preferably 12 per sample. For example, if the DSO has a response extending from d.c. to 500 MHz, then the sampling rate must be at least 1 GS/s (Giga samples per second). At 10 bits per sample this represents a bit rate of 10 Gbit/s. Because of these features, the DSO has a greater accuracy for both the Y system and the X time base.

- A complex signal with many components such as that found in television systems, is not always easy to resolve using either an ART or a DSO.

Because of the extensive add-ons that have been introduced into both systems over time, it has become rather difficult to learn how to manage such instruments effectively with a wide range of features. Practice therefore is particularly important.

The introduction of the DPO by Tektronix Inc, whilst complex in appearance, combines many of the advantages of the ART and the DSO, and in addition provides very much more information about the signals being examined. Apart from data storage and communications over a range of networks, the DPO provides 3-D statistical information via the intensity graduations of the X/Y trace, together with extra on screen information associated with the displayed signal.

Question 5.4

What type of oscilloscope is best suited for tracing timing problems in a microprocessor circuit?

Servicing CROs

By the very nature of the instrument, the CRO carries a great deal of information about its serviceability within its display. Hence by observation of the screen display, many of the typical faults may be localised to one of three separate areas:

- faults that affect the CRT and the power supplies,
- faults associated with the X input and timebase stages,
- faults in the Y amplifier stages that affect the work signal being displayed.

Individual faults can then be located using a second CRO.

CRT replacement

The procedure for replacing a faulty or damaged tube varies somewhat from one oscilloscope to another, and the manufacturers handbook should always be consulted. The glassware is most vulnerable at the tube neck and base connector where the glass changes shape most rapidly.

The following general notes will nevertheless be useful.

1. Always wear safety goggles and gloves to avoid injury from flying glass to avoid the risk of implosion.
2. Cover the workbench surface with a blanket or similar material. Place the CRO on a clean bench, with sufficient space available to lay the CRT alongside it once it has been removed.

3. Disconnect the instrument from the power supply, and discharge all capacitors. Any outer coating of carbon on the tube should be treated with respect, it may have acquired a significant charge which needs to be earthed for safety.

4. Remove carefully the tube base, the high voltage connectors, and the leads to the deflection system. For a circular tube it will be necessary to note the particular alignment in respect of the deflection connectors.

5. Carefully remove the tube clamps and screens. Lift the tube out carefully to protect it from any impact. Place the tube on the blanket so that it cannot roll off on to the floor (circular tubes only).

6. Observe all safety precautions until the displaced tube is safely packed into its crate for disposal and then complete the reassembly of the CRO.

7. Finally test and recalibrate.

Answers to questions

5.1 The trace may be magnetically deflected if the magnetic fields are strong enough to penetrate the magnetic shield around the tube.

5.2 PDA, post deflection acceleration.

5.3 Different wave inputs to both X- and Y-plates.

5.4 DSO, digital storage oscilloscope

6

Test and measurement

Transistors and field effect devices (FETs)

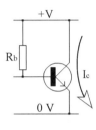

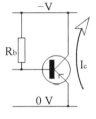

Semiconductor devices were described in Chapter 3 to which the reader is referred for any necessary revision. To recap briefly, transistors are bipolar devices because both electrons and holes (negative and positive carriers) control the action of the device. These devices are sometimes referred to as bipolar junction transistors (BJTs). Such transistors effectively consist of a pair of diodes fabricated back-to-back, in series opposition with the common centre point connected to provide the base control electrode. The two ends of the structure are referred to as the emitter and collector electrodes. In operation, the emitter to collector current is controlled by the base to emitter voltage or current.

From this description it can be seen that the first line of testing can use the concept of the forward and reverse biased resistance for diodes. Using an ohmmeter connected alternatively between collector and base and then between base and emitter, a high or low resistance should be found when reversing the polarity of the test leads. A similar test between emitter and collector will always find one diode forward biased and the other reverse biased so that a fairly high value of resistance should be found with each polarity of test. However, this limited test will only indicate that there are no short or open circuits within the device. Activity 3.2, Chapter 3, explains further how the gain of the BJT can be checked. It will be recalled that there are two variants of the BJT, NPN and PNP versions for which the sketch, left, shows the necessary conditions for conduction of each. Certain BJTs are fabricated as NPN and PNP transistors with virtually identical features and parameters so that they may be used in push-pull circuits.

By comparison, the FET is fabricated around a current carrying channel connected between the source and drain electrodes. Control of the unipolar current flow which consists either of electrons or holes only is obtained via a capacitive effect between a gate electrode and the base substrate. Activity 3.3, Chapter 3, explains further how the gain of FETs can be assessed.

Whilst junction field effect transistors are used in some applications, the more common device in this area is the insulated gate FET. This is also known as the MOSFET (Metal oxide semiconductor FET) where a layer of silicon dioxide is used to form the insulation between the metal layer (usually aluminium) that forms the control gate.

The input resistance can be very high, up to 10^6M in parallel with a capacitance as low as 1 pF.

Do not make resistance checks on a MOSFET using an ohmmeter because the probe currents can easily destroy the gate insulation and the test leads might well act as aerials to induce destructive electrostatic energy from the air.

Looking first at the N-channel MOSFET, illustrated (a) left, with a positive voltage applied to the gate (V_g), electrons are attracted to the

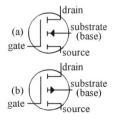

surface of the P type substrate (or base – B) so that an induced N channel is formed beneath the gate that allows current to flow between source and drain (I_d). I_d now increases in a near linear manner with an increase in V_g to provide the so called *enhancement mode*. The complimentary P type version is shown in illustration (b), with the arrow pointing out from the substrate connection.

Alternatively, an N channel can be formed on the surface of the P type substrate during fabrication to provide a conduction path even with $V_g = 0$. Since an increasing negative gate bias now reduces the drain current I_d, this is described as the *depletion mode*. With this type shown, left, (a), a positive gate voltage will increase the conductivity so that this FET can function in both depletion and enhancement modes. The complimentary version shown (b) is fabricated with a P type channel diffused on to an N type substrate.

The substrate (base) can be used to provide a second input terminal but since this has a lower sensitivity to the current flow, the base is more usually connected to the source either internally or externally.

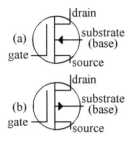

Failure mechanisms of semiconductor devices

Just as diodes can become open or short circuit so similar faults can arise with BJTs between base and collector and base and emitter. Individual electrodes can become open circuit and a condition known as *punch-through* can occur where a short circuit develops between collector and emitter when the base region is ruptured usually through excessive current and hence heat dissipation. Other faults that may be encountered are changes in input and output impedances and gain. Generally these are due to overdriving and the consequent dissipation of excessive heat.

SWA = Safe working area

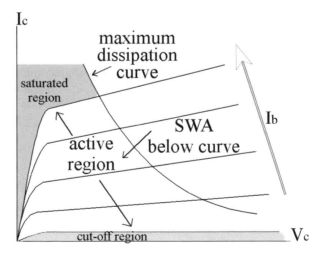

Figure 6.1   Output characteristic for a BJT showing the various areas of operation

For FETs the insulation layer of silicon dioxide between the gate and substrate is the major weakness. The gate region is susceptible to damage through electrostatic discharges (ESD) and for this reason, this electrode is often protected by diodes built into the structure during fabrication to provide a shunt path for such discharges.

Whilst Figure 6.1 shows the output characteristic for a BJT, it is also fairly representative of the FET. At a zero value of base current I_b, the collector current I_c is practically zero for all normal values of collector voltage V_c. This is described as the *cut-off region*. As I_b and V_c increase so does I_c as shown and this represents the *active region* that is used for linear applications. At low values of V_c the collector current lies in the *saturated region* irrespective of the values for I_b. Since both saturation and cut-off regions represent areas of either practically zero voltage or current, the transistor dissipates very little heat and these are the regions used for digital operations. However, it is important that the transition between on and off should be rapid to avoid generating heat and wasting power source energy. This diagram also shows a maximum dissipation curve that describes boundary of the safe working level of power (SWA) that the device can dissipate. It is a curve drawn for all value of maximum power dissipation = $V_c \times I_c$.

Question 6.1

Why are there four varieties of MOSFET and only two varieties of BJT?

Testing semi-conductor devices

Semiconductor devices are relatively easy to test when out of circuit. However, when included on a populated circuit board, the dynamic effect of any signals and the shunting effect of bias and other components give rise to misleading results.

Many modern digital voltmeters (DVM) now include sockets for testing the gain (h_{fe}) of small signal transistors and the operation of this is based on the circuit similar to that used for Activity 3.2. However for other devices such as power transistors, although the basic circuit can be adapted it must be remembered that whilst small signal transistors usually have a gain that is measured in hundreds, the gain of larger power devices will be in the range of tens.

Rather costly dedicated transistor testers are available that can measure the parameters whilst the devices are in-circuit. In use, the circuit carrying the transistor to be tested is switched off, the test unit is connected to the transistor via short low resistance leads and the residual effect of parallel circuit components is backed-off with a balance control. Operating a *push to test* switch causes the meter to indicate the value of h_{fe}.

More sophisticated test instruments are available that can also measure the other important device parameters such as input and output impedances.

Atlas™ tester

The Atlas semiconductor tester Model DCA55 made by Peak Electronics Design Ltd, Buxton, Derbyshire, SK17 9JL is a small palm sized instrument that is capable of testing and identifying out of circuit, practically all types of transistors (BJTs, compound transistors, transistors with shunt diodes or resistors, FETs, MOSFETs, both enhancement and depletion types, etc.), including diodes, diode arrays and LEDs with 2 or 3 leads. Each device can be connected to the tester via three colour coded leads terminated in gold plated test clips without the need to identify the pin-outs. For two-leaded devices any pair of test leads can be used either way round.

The results of the tests including pin identification are displayed on a two line LCD which can be scrolled through to display all the other test data. This includes $h_{f}e$ values ranging from 4 to 65,000, diode and transistor threshold voltages, device volts drop, collector and base test currents, type of device (germanium or silicon, PNP or NPN etc.). The features also make the instrument very useful for selecting and matching the characteristics of diodes and transistors for constructional projects.

This intelligent instrument is controlled via two push buttons, one to start the automatic test sequence and the other to scroll through the display pages. If the instrument is inactive for 30 seconds, it automatically switches off. Alternatively, two quick presses on the scroll button has the same effect.

Power is provided by a 12 volt battery (GP23A) of the type commonly used in car alarm key fobs with a lifetime of about 1 year. The instrument has a built in self test facility with indication of low battery state, all for the price of £60. Due to the relatively large number of shunt components on a circuit board, the results from in-circuit testing are unreliable.

Activity 6.1

From a selection of BJT devices, some of which may be faulty, carry out forward and reverse resistance tests on the three termination pairs. Record the findings and compare the results with further tests carried out using a proprietary transistor tester.

Testing R, L, and C components

All discrete components suffer from ageing and heating effects and, since the latter are particularly cumulative, it is important to avoid excessive temperatures during construction and repair. Furthermore it is important to avoid the hazards of solder fumes and the environmental effects of the lead content of solder. Typically a tin/lead solder with a melting point of about 183°C has been commonly used but this is progressively being replaced by lead-free alloys obtained from metals such as tin, copper, silver, indium and zinc. Like active devices, R, L, and C components are all more difficult to test in-circuit than as individual items. Hence the simple solution is often to unsolder one end of the suspect device. All of these devices can be tested using a universal bridge tester.

Resistors

These components very rarely develop low resistance values in use. They commonly go high in value or develop an open circuit due to age or through passing excessive current and hence heat dissipation. Although front line testing may usefully employ a DMM switched to the ohms range, for the most accurate assessment of change of value, a Wheatstone bridge tester is by far the most effective. (See Chapter 4).

Inductors

These devices may develop open circuits through passing excessive current or simply by corrosion forming at the terminations. More commonly though, inductors. like transformers, develop short circuited turns which dissipate energy. This generates excessive heat and generally adds to the lossiness of the component. Even a short circuit between two adjacent turns can lead to the faulty operation of any circuit to which the inductor is connected.

Capacitors

Capacitors are particularly sensitive to temperature effects and can develop into open or short circuits, change of value or develop excessive leakage across the dielectric. Electrolytic capacitors are most susceptible to this problem. The important property is the effective series resistance (ESR) which can be measured on an instrument similar to a DMM (ESR can even be included as one of the special DMM functions). Generally, ESR is very low, usually less than 1R and represents the loss in phase angle between the current and applied voltage which in theory should be 90°. Capacitors are commonly rated for operations at 85°C, 105°C or 125°C so that if repeated failures occur, the replacement capacitor temperature rating can usefully be upgraded.

Universal bridges

The basic Wheatstone bridge circuit which was introduced in Chapter 4 to determine the values of unknown resistances can be extended to evaluate unknown inductors and capacitors. In this case, the bridge must be energized from an a.c. source, the unknown device must be compared with a similar standard component, and the null detector must include a suitable a.c. rectifier circuit.

The basic conditions for the bridge circuit at balance are that the products of opposite arms are equal. In the universal bridge, the product of the range and balance resistor values must be equal, at balance, to the product of the unknown impedance and the standard impedance.

The simple bridge can only determine the component value and not evaluate its losses. In general, it is also necessary to determine the value of the equivalent series resistance of an inductor and the equivalent shunt resistance of a capacitor. In both cases, the resistive element represents the component losses which are related to its quality factor; the Q factor in the case of the inductor, and the loss angle (the angle by which the voltage/current relationship fails to reach the theoretical 90°) for the capacitor.

Figure 6.2 shows how the basic bridge concept can be adapted to provide universal features. This bridge circuit may be energized either from an internal 1 kHz oscillator or from an external source. Using an external

10 kHz source is particularly useful for measuring components with very low losses.

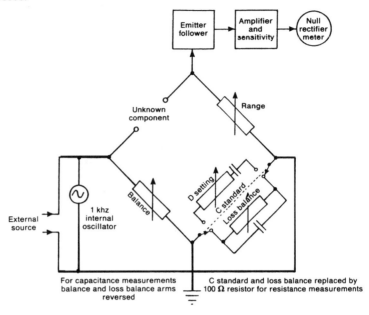

Figure 6.2   Universal bridge (courtesy of Marconi Instruments Ltd.)

The null detector circuit and its amplifier and sensitivity control are buffered by an emitter-follower circuit to minimize loading on the bridge network. In application, the unknown component is connected into the appropriate arm of the bridge. The instrument is then set to measure either an R, L, or C component and a range is selected that provides a minimum meter deflection. A lower minimum value is then obtained by adjustment of the balance control and this then indicates the component value. By further adjustment of the loss balance control, a true zero meter indication can be achieved. The loss balance setting then indicates the resistive loss component. The range of measurements available with this instrument, with an accuracy of better than ±1 per cent of the reading obtained, is indicated here, left.

Resistance: 0.0R to 11M

Inductance: 0.2 µH to 110 H

Capacitance: 0.5 pF to 1100 µF

Servicing of universal bridges

Certain of these instruments are designed for portable field work and are therefore likely to be subjected to shock and vibration. Local service work is therefore likely to involve the repair of such ravages. Since the overall accuracy is critical, any repair work must be followed by re-calibration. The necessary equipment for this is only likely to be found in specially setup service departments.

Activity 6.2

Using a universal bridge, compare the values obtained for a number of R, L and C components each with nominally the same values. In particular, inductors and capacitors can show a marked difference in loss values. It can also be useful to measure the self-inductance of some wire-wound resistors.

Question 6.2

What instrument would you use to make accurate measurements on values of passive components?

Counter/timer

These devices, which range from simple hand-held models to complex laboratory bench instruments, are chiefly used for measuring signal frequency (or pulse-repetition frequency) over a range, typically from d.c. to more than 2 GHz.

The operation is based on pulse-counting over a known time period, from which the frequency is automatically determined. The read-out is then commonly presented on a seven-segment LCD or LED display. The block diagram of the basic system, omitting the reset and synchronizing sections is shown in Figure 6.3.

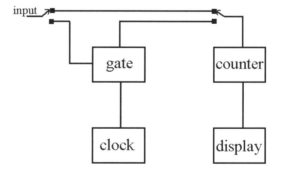

Figure 6.3   Basic block diagram of a counter/timer

The counter section consists of a binary counter with as many stages as required for the maximum count value. The input to the counter is switched, and in the counting position, the input pulses operate the counter directly. When the switch is set to the timing position, the input pulses are used to gate the clock pulses, and it is the clock pulses that are counted.

Counter/timers are often microprocessor-controlled with a basic clock frequency, typically of 1 MHz. This is provided either by a temperature-compensated crystal oscillator (TCXO) or by an oven-controlled crystal oscillator (OCXO). Provision is often made for two-channel (A and B) measurements and these inputs may have different upper cut-off frequencies. Low-frequency noise can be troublesome with these instruments and so a low-pass filter with a cut-off frequency at about 10 kHz is often used at the inputs.

Facilities include absolute frequency, phase, period and time measurements, with frequency and time ratios for the A/B channels. Selectable gate times are used to extend the range of measurement. To extend the frequency range even further, a prescaler (a frequency divider ×10, ×100) can be connected between the source and the counter/timer. Typical sensitivities range from about 20 mV at 5 Hz to about 150 mV at 150 MHz, with input impedances of the order of 1M in parallel with 40 pF. As for the CRO, ×10 probes are usually available for use with signals of an amplitude greater than the in-built attenuator can handle. The resolution (the smallest change that can be registered) varies from about 0.1 Hz at the lower frequencies to 1 kHz at HF.

Microprocessor controlled instruments allow for the inclusion of more elaborate features, such as hard-copy printouts and links to automatic test equipment (ATE) networks.

Servicing problems

Apart from power-supply problems, all the fault tracing will use digital techniques. Again, it is important that any service work should be concluded by re-calibration, therefore this form of work should only be entrusted to suitably equipped service departments.

Frequency counter

The frequency meter makes use of a high-stability crystal controlled oscillator to provide clock pulses. The instrument may either be combined with a counter/timer or provided as a function of a complex DMM, in which case, the accuracy will be somewhat reduced. In its simplest form, (see Figure 6.4), the unknown frequency used to open a gate for the clock pulses.

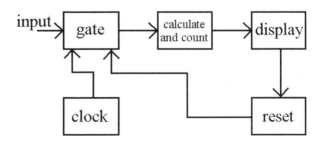

Figure 6.4   Block diagram of a basic frequency counter

The number of clock pulses passing through this gate in the period of one input cycle will provide a measure of the input frequency as a fraction of the clock rate. For example, if the clock rate is 10 MHz and 25 cycles of the unknown frequency pass the gate, then the input frequency is 10/25 MHz, which is 400 kHz. A calculating and counter circuit then passes this result to the display stage to give a direct indication of the input frequency.

In this simple form, the frequency meter cannot cope with input frequencies that are greater than the clock rate, nor with input frequencies that would be irregular sub-multiples. For example, it cannot cope with 3.7 gated pulses in the time of a clock cycle. Both of these problems can be solved by more advanced designs.

The problem of high frequencies can be resolved by using a switch for the master frequencies that allows harmonics of the crystal oscillator to be used. The problem of difficult multiples can be solved by counting both the master clock pulses and the input pulses, and operating the gate only when an input pulse and a clock pulse coincide. The frequency can then be found using the ratio of the number of clock pulses to the number of input pulses.

Suppose, for example, that the gate opened for 7 input pulses and in this time passed 24 clock pulses at 10 MHz. The unknown frequency is then $10 \times 7/24$ MHz which is 2.9166 MHz. Frequency meters can be as precise as their master clock, so that the crystal control of the master oscillator determines the precision of measurements.

The instrument is useful for checking and adjusting a wide range of oscillator settings, crystal oscillators in frequency conversion stages, display driver clocks and motor speed controllers.

The read out should extend to at least 8 and preferably 10 digits with an accuracy of at least 2 ppm (parts per million – 2 in 10^6) and with a sensitivity of at least 10 mV. A prescaler can be used to extend the range of frequency operation and pre-selectable gate times of 0.1 s, 1 s and 10 s are often provided, but with progressively longer settling times.

For re-calibration purposes, a service accessible preset adjustment provides means of checking the master oscillator setting against some frequency standard. In the absence of a suitable laboratory standard use can be made of the 50 Hz and 15.625 kHz field and line frequencies from a TV receiver, the broadcast colour TV sub-carrier at 4.43361875 MHz or even the BBC Long Wave programme at 198 kHz whose carrier is maintained to an International Frequency Standard.

Low frequency signal generator

By comparison with other signal sources, these instruments are relatively simple devices. As shown in Figure 6.5, the fundamental frequency is generated typically by a Wien bridge oscillator stage because of its wide basic frequency span. This stage is followed by a buffer amplifier to isolate the oscillator from the effects of any load. The output signal level is controlled by an attenuator stage that delivers an output typically at 600R impedance.

Low frequency signal generators are usually employed for the setting up and servicing of audio systems and as such they produce output signals of both sine and square wave form. The frequency range usually covers from 10 Hz to 1 MHz in five switched decade bands, with a frequency accuracy ranging from ±1 per cent to ±5 per cent at full scale.

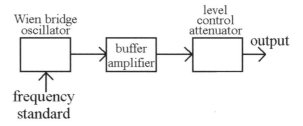

Figure 6.5 Block diagram of a low frequency signal generator

Because of the needs of hi-fi systems, these generators usually provide a very low level of sine wave distortion, typically better than 0.05 per cent over the range 500 Hz to 50 kHz. Off-load outputs of 20 V p–p are common with the level being controlled in a series of steps of –10 or –20 dB, down to about 200 mV minimum.

Question 6.3

How would you check the calibration of a frequency meter in the absence of a standard frequency generator?

Activity 6.3

Measure the stage gains of a 3-stage audio amplifier using the controlled output of a low frequency signal generator, together with a suitable instrument to measure the signal amplitudes. Evaluate the –3dB cut-off frequencies to determine the bandwidth for the complete amplifier.

Function generators

Basically these are signal generators designed to produce sine, square or triangular waves as outputs. Each may be varied typically over a frequency range covering from less than 1 Hz to about 20 MHz. The basic frequency may be generated either by a highly stable oscillator circuit, or by using a frequency synthesizer. The output levels are typically variable between about 5 mV and 20 V peak of either polarity, plus a TTL-compatible signal at ±5 V. Some of the basic waveforms are shown in Figure 6.6. The output impedance is nominally 600R and provision is made for driving signals via balanced or unbalanced lines.

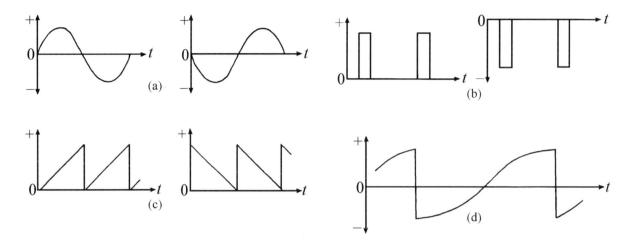

Figure 6.6　Function generator waveforms (a) sinusoidal, (b) positive and negative pulses, (c) positive and negative sawtooths (serrasoids), (d) compound wave

Each of the basic output signals is capable of being modified. The sine wave may often be phase-shifted, and the square wave mark-to-space ratio is variable so that a pulse stream of variable-duty cycles can be provided. The triangular wave can be varied to provide a sawtooth of varying rise and fall periods. In addition, it is also possible to add a d.c. value to each output as an offset. This is valuable for testing circuits that are d.c.-coupled and with a frequency response that extends down to zero. The square wave is also differentiated and integrated to generate exponential envelopes (see Figure 6.7).

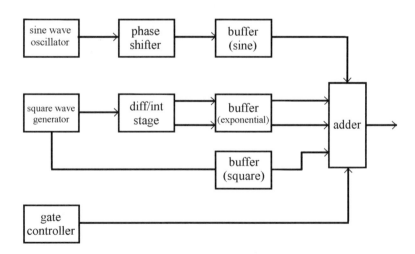

Figure 6.7　Block diagram of function generator

An additional feature often allows a second signal component to be gated into the primary waveform in the manner indicated in Figure 6.6(d). A further variation of this is the *tone burst* that consists of a sine wave modulated with a parabolic envelope, but generated in bursts. This waveform is particularly suited to the testing of full-bandwidth audio systems or those circuits where strange resonances may occur. A variation of this feature utilizes a form of FM. Here, a basic sinusoid is frequency-swept between predetermined limits, either at a linear or logarithmic rate.

The more sophisticated instruments in this range may be equipped for microprocessor control, have a digital read-out of frequency and amplitude, plus a built-in standard frequency source.

Servicing problems

It is most important that the instrument calibration is carefully checked following any repairs. Service work is therefore a very specialized operation. Integrated circuits are extensively used, together with stabilized switched-mode power supplies. The highly critical oscillator circuit is invariably temperature controlled.

Activity 6.4

With a function generator providing the input signals, practise using an oscilloscope in all its modes, to gain experience of using both instruments. Examine all the possible waveforms that the function generator can produce.

Pulse generators and detectors
Logic probes and pulsers

Logic probes and pulsers are complementary devices, with the pulser acting as a signal generator and the probe as a signal tracer. Both are small hand-held devices which are powered from the circuit under test. They are therefore ideal front-line field service test instruments for digital circuitry.

The probe is used to test the signal state along bus lines or at component inputs and outputs and it provides the state indication on LEDs. In a complementary way, the pulser provides an output square wave that is injected into strategic points in the circuit under test so that the outputs can be monitored either on a logic probe or using a CRO.

For more details of logic pulsers and probes, see Chapter 16.

Used in conjunction with the service manual and/or circuit diagram, these instruments can be used to locate a significant number of fault sources.

Question 6.4

What is the simplest instrument available for checking logic states of digital circuits?

Signal level measurements using decibels

We saw in Chapter 4 that it was convenient to express sound intensity levels using the logarithmic scale of decibels. This concept can usefully be applied to the measurement of stage gains and losses in communications networks.

Example: if a correctly matched transducer provides an input of 2 dBm (2 dB above the level of 1 mW), then to provide an output of 2 watts, the amplifier needs to have a gain of 31 dB.

Solution: 2W = 2000 mW or in dBm = 10 log 2000 = 10 × 3.3 = 33 dBm.

Input level = 2 dBm, so that amplifier gain = 33 – 2 = 31 dB.

Example: A 3 stage amplifier includes a complex tone control circuit that has an attenuation of 0.4. The individual voltage gains of the active amplifiers is G_1 = 30, G_2 = 25, and G_3 = 10.

Solution: The overall system gain can be calculated as follows;

G_1 = 20 log 30 = 29.5 dB

G_2 = 20 log 25 = 28 dB

G_3 = 20 log 10 = 20 dB

A_1 = 20 log 0.4 = –8 dB.

The overall gain is thus. 29.5 + 28 + 20 +(–8) = 69.5 dB.

Noise in a communications system can be considered as a destroyer of information and thus needs to be quantified in some way. Normally this is described as the system signal to noise ratio (SNR or S/N), which is expressed as

SNR = 10 log (signal power/noise power) dB, or alternatively as

20 log (signal voltage/noise voltage) dB (or the similar expression relating to current when this is more appropriate).

All components in an electronic system that pass a current generate a small level of noise due to the random motion of the electrons. Thus an amplifier or receiver system will add further noise to its input signal to degrade the SNR further.

For example, if a receiver has an input signal of 500 µV together with 25 µV of noise superimposed from the transmission medium, the input

SNR = 20 log (500/25) dB = 20 log 20 = 26 dB.

If the receiver has a processing gain of 10 the output signal due to the input alone = 5 V + 250 µV, so that the SNR is still equal to 26 dB.

However if the receiver adds a further 50 µV of noise, the total noise level at the output becomes 250 + 50 = 300 µV. The overall SNR now becomes 20 log (5000/300) = 20 log 16.6666 = 24.5 dB. Thus the receiver noise degrades the SNR by 1.6 dB.

In the analogue domain, noise simply reduces the intelligibility of the signal making it difficult to understand or interpret the meaning of the information. However in the digital domain, noise leads to the corruption of data bits to produce a *bit error rate* (BER) but the overall effect is still the same.

Activity 6.5

Using a standard VHF/FM receiver, terminate the aerial input socket with a 75R resistor, turn the volume control to maximum and with a suitable meter connected across the loudspeaker terminals, measure the noise output. Remove the input resistor, tune the receiver to a station and again measure the sound output level. Calculate the receiver SNR.

Answers to questions

6.1   MOSFETs can be N or P channel, and operated in depletion or enhancement mode. BJTs are either PNP or NPN.

6.2   Universal bridge.

6.3   Use broadcast TV signals or radio long wave.

6.4   Logic probe.

7 Using the PC

Servicing in the modern environment

Due to the introduction of computing systems, the lifetime of the hard copy service manual is nearly over. Much of today's service information is now either provided on CD-ROM or via the Internet. In earlier times, the service technicians manual has not only carried the original service diagram and data, but also very much personal information that he/she has gleaned over time about that particular system. Trying to read today's manual from a monitor screen is certainly a different experience. Following the diagram of such as a colour TV receiver with interactions across many screens, appears at first sight virtually an impossibility. Even by printing out sections of the diagram and gluing them together with sticky tape is not conducive to best use of the servicing bench. Fortunately, most electronic based service information provides space in memory for the technician to add further comments as aids to memory for future use. However the service information is provided, either via the Web, CD, VCR tapes, DVD, or even satellite links, new skills will have to be developed to make the most of what could be information overload.

Workshop support databases

Technicians servicing domestic electronic equipment are generally well provided with circuit diagrams and data sheets, but to produce a thoroughly reliable and economical repair that will generate customer loyalty, they need additional skills and support. Not only is it necessary to understand how the system works, there is also a need to have a knowledge of the system's historical reliability and its particular points of failure. In the past when systems were constructed largely from discrete components, many of the failures occurred in a regular manner, giving rise to the *stock faults* Indeed, many service departments found these to be a source of good business that created a sound reputation for doing a good job.

With today's extensive use of dedicated ICs the system reliability has improved very considerably. However, stock faults still occur but repeat very much less frequently making a good memory an additional requirement for service personnel. One of the best service aids is a subscription to the journal *Television* (see below). This acts very much as a clearing house for the hints, tips and solutions to problems encountered by many practising service technicians. Fortunately, there are now a number of computer system databases available that have been designed to aid the servicing of such equipment. Whilst these can provide almost instant access to many of the stock faults, it must be emphasized that these should be used in conjunction with the manufacturer's circuit diagrams and data sheets. Of these databases, two have been tested and found to be invaluable to the busy workshop.

The first one tested was provided on CD-ROM by SoftCopy Ltd, Electronic Publishers, Cheltenham, Gloucester, GL53 0NU. This has been produced in co-operation with the journal *Television*, now published by Cumulus Business Media, and is updated annually. The program has been designed to run on a relatively low level IBM compatible personal computer, contains almost 15000 different entries, more than 200 equipment specific servicing articles and is very easy to use. There is additional space within the system memory to incorporate the results of personal experience.

The database covers faults on CD players, DVD players, camcorders, satellite TV systems, VCR machines, computer monitors as well as television receivers that have been published in the journal back to 1988. In addition, there is a directory reference to many of the published articles that date back as far 1986. These provide good background reading to obtain an understanding of the operations of many of the popular items of equipment. It is thus a tool that should find a slot in any repair workshop. Because of its relatively low processing needs, this database could also be installed on a lap-top computer for use by the field servicing engineers.

The database provided for testing by EURAS International Ltd, Keynsham, Bristol, BS18 2BR, is also supplied on a CD-ROM and to some extent this increases processing needs. However, this is updated three times annually on a subscription basis and covers the same wide range of domestic entertainment and computer equipment. The fault finding and repair hints represent the knowledge gleaned from manufacturers, dealers and repair centres, and covers about 300 000 entries from more than 400 manufacturers. Unusually, the database information is covered in most European languages. Again the system memory provides space to record personal knowledge gained over a period of time. Because of the greater available memory space on the CD-ROM, the publishers can provide the appropriate circuit diagram segments with each repair hint.

Some useful servicing related web sites for technical data, service sheets, spares and data are listed at the end of this chapter.

Question 7.1

What are the main alternative options to the use of service manuals?

Activity 7.1

Use a PC based system to access service and spare parts data from both a CD-ROM and a number of web sites.

Virtual instruments (VI)

With the introduction of the PC to the workshop environment, a wide range of hardware devices that are software controlled have become available. These range from fairly simple and low cost active probes capable of providing storage oscilloscope, spectrum analyser, digital voltmeter

(DVM) and even transient recorder functions, through digital multimeter (DMM), storage scope with spectrum analyser, transient recorder, plus function generator, to high cost extensive networked systems that provide all of the above, plus remote data logging from test benches etc.

Generally these all have one point in common, they are attached to the PC via the parallel printer port, a specially provided plug in parallel port card, or, more recently, the USB (universal serial bus) connector.

Question 7.2

What is the main advantage of using an Internet search engine to solve a technical servicing problem?

VI suppliers

Although there are other suppliers of VI equipment, only the following devices and systems that run in the familiar Windows environment with typical task bars and pull-down menus, have been evaluated for the servicing environment. These are manufactured by:

National Instruments Corporation (UK) Ltd. www.ni.com/uk,

Pico Technology Ltd, www.picotech.com and

TiePie (Engineering (UK), www.tiepie.com.

National Instruments

Of the three manufacturers, NI probably provides the most extensive range of measurement, data acquisition (DAQ), data logging including image acquisition, and signal conditioning accessories for a system that is more likely to be used in a product life testing and monitoring environment. Since the system is designed to operate with real time applications, in a measurement, control and statistical analysis mode, it is capable of being coupled into a wide range of communications networks that include, back-plane bus extensions, Internet, GPIB, Ethernet, Firewire (IEEE1394) and USB (universal serial bus).

Standard function generator waveforms including composite video are provided plus many more that are user programmable, operating at up to 40 MS/s with 16 bit resolution. Up to eight counter/timers which are 5 V TTL compatible, can operate at up to 80 MHz with a resolution of 32 bits.

The digital multimeters (DMMs) have 5½ digits accuracy for a.c. voltage and current, (true r.m.s.), d.c. voltage and current, and resistance (Ω). These are used to measure the outputs from many different types of transducers on several channels virtually simultaneously.

The data capture analogue to digital converters (ADC) or *digitizers* can be driven from a wide range of transducers via appropriate signal conditioners, ranging from resistance temperature detectors (RTD), thermocouples, thermistors, chromatography sensors, strain gauges, force, load and pressure sensors, linear displacement devices such as the LVDT (linear variable differential transformer). Bandwidths vary from 100 MHz down to 4 MHz whilst the corresponding resolution ranges from 8 bits up to 21 bits.

Because of the complex nature of the NI system, the manufacturer provides extensive technical back-up that ranges from system development support, technical training and future development.

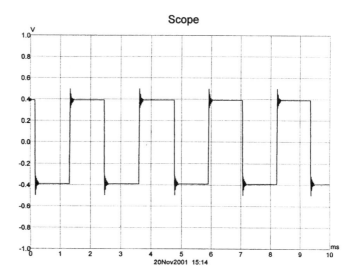

Figure 7.1   CRO display of square wave signal in time domain

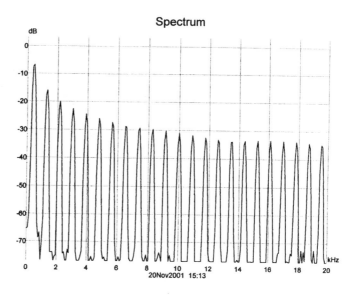

Figure 7.2   CRO display of square wave in frequency domain

Activity 7.2

The waveform shown in Figure 7.1 can be used to derive the relationship between the peak and RMS value for a symmetrical square wave as follows. Squaring the voltage waveform that swings between −0.4 and +0.4 volts, produces a new wave for V^2 as follows:

$(−0.4)^2 = +0.16$ and $(+0.4)^2 = +0.16$.

The average value for V^2 is thus 0.16 and continuously positive. Taking the square root of V^2 produces the values of 0.4 volts again continuously positive. Thus for the symmetrical square wave the peak and RMS values are equal. This is only true for square waves, all other shapes have a variation in this relationship. Now suppose that the addition of a d.c. offset produces an asymmetrical square waveform that swings between −0.2 volts and +0.6 volts (amplitude is unchanged). Repeat this exercise and compare the results.

Pico Technology Ltd.

This equipment provides similar types of functions but with lower accuracy and less sophistication and at lower cost than the NI system. The Pico system provides an economical alternative to standard test equipment and data acquisition tools. The system software and documentation is supplied in the most European languages, with a back-up service via the Internet and Web site. The system is fully developed to CE under ISO9001 standards. It functions on most of the early PC developments from DOS software and 80386 or higher processors supported by downloadable updates from the Website. The adapter plugs into PC parallel port from which it derives its power supply.

In the time domain, the main screen of the oscilloscope provides for both normal timebase and X–Y operations. The latter is particularly useful when comparing two time related waveforms such as found with advanced modulation schemes. The spectrum analyser provides an alternative view of a signal in the frequency domain (see Figures 7.1 and 7.2). Switchable probes of ×1/×10 or ×10/×100 are available.

In the digital storage CRO mode, the sampling speeds range from 20 MS/s up to 100 MS/s with comparable analogue bandwidths. Both waveforms and instrument settings can be stored on disk for future study and analysis.

Plug in cards are available to add extra PC parallel ports and/or a USB adapter.

0.392V

431Hz

The DMMs provide for 1 to 22 channels of inputs with resolution ranging from 8 to 16½ bits (sign bit). Two of its typical readings are shown, left. The sampling rate range from 20 kS/s at 8 bits for the single channel down to 10 kS/s at 10 bits for 22 channel operations. The input impedance varies from 200K for a single channel up to more than 1M for multichannels.

When equipped with suitable sensors/transducers, the data acquisition and logging software can measure temperature, humidity, sound pressure, light, current, resistance, power, speed and vibration levels. This data can be transferred to other Windows applications over local area networks or the Internet using the *copy and paste* facility. The data can also be displayed in a spreadsheet format, such as Excel, with analysis of data trends.

The Pico system can be operated from a laptop PC via the PCMCIA slot and a rechargeable battery pack. Field operation can be extended by an adapter power supply lead that plugs into a car cigar lighter socket.

The system described can be extended to provide an environmental monitoring network with remote logging, sensors and alarms that include telephone autodialler, modem adapter and many other useful features. In addition, many useful educational experiments are included in the system library.

TiePie (Engineering) Ltd

This virtual instrument system consists chiefly of two quite separate devices together with a range of plug-in PC interfaces and probes. The small *Handyprobe* is as its name suggests, is a small hand-held probe-like instrument that can be used to probe around a circuit board or system to obtain data about its signal conditions. This device is directly coupled to the PC parallel port from which it draws its power supply.

Applications software then converts the PC into a storage oscilloscope, spectrum analyser, voltmeter with a wide range of functions and a transient recorder. The software provides an intelligent auto-setup whereby the device automatically adopts an overview of any signal presented via the probe. Point and click by mouse control then selects the user required waveform information. All previous applications settings can be saved to disk memory for future use, thus reducing setup time to a minimum. Cursor control can select any one of 21 possible read-outs.

Text balloons and text lines are provided to highlight any particular signal points that may require further study after being saved to disk. The probe has an 8 bit resolution with a maximum sampling rate of 20 MS/s, together with an input range that varies from 0.5 volts up to a maximum of 400 volts full scale, with an accuracy of 1 per cent ± 1 LSB and an input impedance of 1M in parallel with 30 pF. It runs on anything above a 486 processor with Windows/DOS software from the early 1990s onwards. It requires only 8 Mbytes of RAM and uses record lengths of 32 kBytes / 64 kBytes.

Oscilloscope mode: The single channel mode provides a bandwidth of 2 MHz with a maximum sample rate 20 MS/s. The timebase ranges from 2 μsec/div to 6 sec/div and provides for a wide range of triggering features.

Spectrum analyser mode: The frequency range is from 0.1 Hz to 10 MHz, with an accuracy better than 1 per cent. The vertical axis (amplitude) is linear/dB with the horizontal axis (frequency) being either linear or in 1/3 octave bands. The Handyprobe in this mode also has a significant number of built in mathematical functions for such a small instrument.

Voltmeter: True RMS voltmeter functions are provided that cover the range from 10 Hz to 5 MHz, with up to 3 user selectable displays covering up to 11 mathematically derived parameters.

Transient recorder mode: This provides for up to 32 768 measuring points over a time scale ranging from 0.01 seconds up to 300 seconds.

TiePieScope HS801:

This unit has all the features of the Handyprobe but with a very much higher specification. It is enclosed in a small case, with provision for mains power supply at 90–260 V a.c. or 12–24 V d.c. for portable applications. Like the Handyprobe it operates with the familiar Windows task bars and pull-down menus and requires only a relatively low level of PC running anything from Windows 3.x and DOS 5.1 onwards for operating systems. It is provided with 2 (×1 / ×10) switched probes and complete instruction manual. Once the software has been installed, it is immediately ready for operations.

The basic instrument provides storage oscilloscope, spectrum analyser, multimeter and transient recorder functions. An upgraded version (HS801-AWG) provides an additional arbitrary waveform generator that operates as a function generator. The input impedance is again 1M in parallel with 30 pF.

Text balloons can be added to highlight specific points about each signal display plus three lines of text to provide future references after storing data on disk. This instrument has the same quick setup and save features that were found in the Handyprobe, with all setting files saveable to disk for future applications.

Oscilloscope mode: The bandwidth is 50 MHz with a sampling rate ranging from 0.002 S/s to 100 MS/s. The timebase extends from 1 μs/div to 600 s/div with a wide range of triggering options. The measuring modes include Ch1, Ch2, Ch1 + Ch2, Ch1 – Ch2 and Ch2 – Ch1, plus an X–Y mode.

Spectrum analyser mode: The frequency range extends from 0.001 Hz to 50 MHz with an accuracy of better than 0.1 per cent. The analysis is based on 8 bit resolution and with an input ranging from 0.1 volts to 80 volts full scale. The amplitude axis is scaled in linear dB, with the horizontal frequency axis scaled linear, logarithmic octave, or 1/3 octave bands. The instrument is also equipped to handle digital Fast Fourier Transform (FFT) mathematical processing. The distortion calculations are based on from 1 to 100 harmonics with the scaling in dB or percentages.

True RMS voltmeter: This has a frequency response ranging from 10 Hz to 25 MHz and with a display accuracy of 2 per cent ±1 LSB. Up to six simultaneous meter displays are available on a user selectable basis. Read-outs include, true r.m.s., peak-to-peak, mean, maximum, minimum, dBm, power, crest factor, frequency, duty cycle, rise times, slew rates and total harmonic distortion.

Transient recorder mode: Up to 32 768 measuring points spaced by 0.01 seconds to up 500 seconds.

Arbitrary waveform generator (AWG): This additional option with 50R output impedance provides sine, square, triangle and noise waveforms as standard, plus a wide range of user programmable wave shapes, in the frequency range 0 to 2 MHz in 0.01 Hz steps with 10 bit resolution. The output is digitally controlled in 1024 steps for each range up to a maximum amplitude of 12 volts.

Activity 7.3

With a PC virtual instrument, connect the system to a working three-stage low frequency amplifier to measure the frequency response, bandwidth and gain of the individual stages. Using the spectrum analyser and an input from the function generator set to 100 Hz sine and square waves, note the change of the harmonic content at the amplifier output.

Question 7.3

What is the alternative to equipping a workshop with a full range of specialized test and measuring equipment?

Sources of information

Engineering information, BBC	www.bbc.co.uk, www.bbc.co.uk/enginfo
NTL	www.ntl.co.uk
Transmitter information	www.tvtap.mcmail.com (Transmitter alignment program)
Service support database	EURAS International Ltd, www.euras@euras.co.uk
	SoftCopy Ltd, www.softcopy.co.uk
Service information forum	www.E-repair.co.uk
	www.elmwood,guernsey.net/index.html
	www.skyinteractive.net/tec
	www.repairfaq.org/Repair
	www.repairworld.com
	www.ICHE.com
Specialist repair services	www.mces.co.uk
	www.netcentral.co.uk, (Satcure)
Test equipment	www.Test equipmentHQ.com

	www.vanndraper.co.uk,
	www.cooke-int.com
Manufacturers services	www.philips.com
	www.ti.com, (Texas Instruments)
RS Components	http://rswww.com

Answers to questions

7.1   CD-ROM, Internet, VCR, DVD.

7.2   Collected experience of a very large number of reports.

7.3   Use a PC along with virtual instruments.

8 Reliability and SM repairs

Quality and reliability

Quality is often defined as a measure of how well a product or system conforms to its claimed specification. Reliability is then a measure of how long the product or system continues to meet this specification. Quality and reliability are of great interest to the end user because this directly affects his/her confidence that the product or system will meet his/her expectations. These broad parameters can be evaluated from statistical evidence obtained by manufacturers quality assurance (QA) or control programme. Just as an analogue signal can be digitized for processing by sampling and a highly accurate version of the original regenerated afterwards, a statistical sampling scheme can give an accurate assessment of the state of a population.

Statistical data is therefore acquired through inspection and testing, using standard sampling plans. In addition, life testing of randomly selected batches can expand the information in this data base of product statistics. This then can be applied to failure analysis. A further important feature of user confidence is the concept of traceability. That is, component parts of a system have all been produced and tested under a recognized quality control assurance scheme so that lifetime failures of all elements can easily be traced back to the original source scheme. Such information can be used to improve the overall reliability of a product or system. Since the original system design must take such statistics into consideration, a worst case design methodology must be employed.

Basically a QA programme involves testing and inspection of a specified number of components from each batch of a product and rejecting the whole batch if the number of defectives exceeds some level. Since the samples are selected at random, the technique leads to either a consumers or producers risk as each batch has a risk of being wrongly accepted or rejected. Confidence achieved by a QA programme can minimize the cost to both. The number of samples per batch needs to be large enough to give confidence and small enough to be cost effective. All these features are assured if an industry standard scheme is employed.

Accelerated life testing

Normal life testing involves operating systems or components as near as possible to the end user's applications and conditions. However, if the accepted life time is very long, then data gathered by this technique is too time consuming so that the test procedure has to be accelerated. For example, devices designed for intermittent operation would be run continuously, thus compressing the life span. Electronics components that are intended for continuous operation may be tested with an increased load or at an elevated temperature. Further environmental tests which simulate extreme conditions of temperature, humidity, pressure, shock, vibration and abnormally dirty conditions in a time cycled way, might be used.

Similar principles may be employed for so called *soak testing* (see Chapter 7 for the use of VI systems in this application) after equipment has been repaired but the facilities in many smaller repair depots may not run to such elaborate techniques, the use of a freezer spray or hot air blower can be used to simulate unusual conditions; even placing a blanket over the larger systems can be helpful, if judiciously employed.

Bath-tub diagram

This is the jargon term used to describe the time/failure rate curve. The high levels of early failures shown in Figure 8.1 is often described as *infant mortality*. These are usually due to manufacturing faults, deficiencies in the quality control scheme, design deficiencies, and misuse by or the inexperience of the end user. These failures are fairly quickly eradicated and the failure rate settles down to a constant, often zero level over the normal working life time. Towards the end of the useful service life, the failure rate starts to rise once more and this is described as the wear out region.

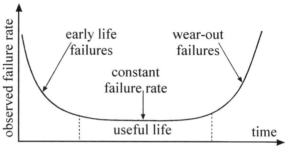

Figure 8.1   The bath-tub diagram.

Mean time between failure (MTBF) or mean time to failure (MTTF)

For a system with a constant failure rate of say 0.001 per hour, it would be expected to fail and have to be serviced every 1000 hours. Thus the average or mean time between failures would be 1000 hours. MTBF is therefore the reciprocal of the constant failure rate. When it is not practicable to replace a failed component, it is more usual to refer to the mean time to failure.

Question 8.1

What does the term *infant mortality* mean as applied to servicing?

Down time or mean time to repair (MTTR)

This is the term used to describe the length of time that a system is out of action between failure and repair.

From the above it can be seen that if components and systems are operated at below their normal design parameters, their life time can be extended, a technique often described as *derating*. In fact capacitors often fail in a system that is running at near to the upper temperature limit for the components. Increasing the rating of a capacitor from say 85°C to 105°C will improve the system reliability.

Even in the servicing environment, it is important to report consistent failures because this can be accumulative throughout a wide area and this could lead the manufacturer to improve the design.

Activity 8.1

Set up a soak test bench using a switched mode power supply circuit board. Connect voltage and current meters to appropriate points and a CRO to the switching device. Monitor the circuit behaviour as first freezer spray, and then a hot air blower, are applied to the board.

Anti-static precautions and handling

MOSFETs, CMOS and Group iii/v (GaAs, etc.) semiconductor devices can easily be damaged by the discharge of static electricity. A discharge from a p.d. as low as 50 volts can cause component degradation and, since the effect is cumulative, repeated small discharges can lead to ultimate failure. It is therefore important that sensitive devices, such as low noise block (LNBs) converters, should be serviced in a workshop where static electricity can be controlled much better than on site. Bipolar transistors (BJTs) are rather more robust in this respect.

The basic work station should provide for operators, work bench, floor mat, test equipment and device under service to be at the same electrical potential. Operators should be connected to the work bench via a wrist strap. Since static electricity is often generated by friction between dissimilar materials the operator's clothing also needs to be considered. The wearing of wool and man-made fibres such as nylon creates considerable static. One very useful garment is a smock made of polyester fabric, interwoven with conductive carbon fibres. This has through-the-cuff earthing. The use of compressed air to clean down circuit boards can actually generate static and this can be avoided by the use of an ionized air blast.

- Static-sensitive components should be stored in conductive film or trays until required and then handled with short-circuited leads until finally in circuit.

- Soldering iron bit potentials can also be troublesome unless adequately earthed.

- The work bench surface should be clean, hard, durable and capable of dissipating any static charge quickly. These static-free properties should not change with handling, cleaning/rubbing or with ambient humidity.

- The use of an ionized air ventilation scheme can be an advantage. Large quantities of negatively and positively charged air molecules can quickly neutralize unwanted static charges over quite a significant area.

Question 8.2

Name the TV device whose servicing is most likely to give rise to electrostatic damage risks.

Surface mount technology (SMT)

This technology is based on a full range of small, leadless components that are available for direct soldering to the metallic conductor pads on the circuit boards. These are smaller than their conventional counterparts and this leads to smaller sub-systems. Because of the relatively large area of the solder pads, highly reliable joints are produced with good heat dissipating properties. The components are more efficient and their self-resistance, capacitance and inductance are much lower, so that these devices have better RF characteristics. The technique is applied to all areas of electronics, from audio through to microwave systems. The technology of surface mounted components or devices (SMC or SMD) is highly compatible with automated assembly, which in turn further improves the cost effectiveness.

Coding of components

Some SMDs are too small to value-code in the conventional way. These should therefore be stored in their packaging until actually needed, to avoid mixing components that are difficult to identify. On the slightly larger components, the coding shown in Table 8.1 is commonly used.

Table 8.1　SMD Component values.

A = 1	B = 1.	C = 1.2	D = 1.3	E = 1.5	F = 1.6	G = 1.8
H = 2	J = 2.2	K = 2.4	L = 2.7	M = 3	N = 3.3	P = 3.6
Q = 3.9	R = 4.3	S = 4.7	T = 5.1	U = 5.6	V = 6.2	W = 6.8
X = 7.5	Y = 8.2	Z = 9.1				
a = 2.5	b = 3.5	d = 4	e = 4.5	f = 5	m = 6	n = 7
t = 8	y = 9					
$0 = \times 1$	$1 = \times 10$	$2 = \times 100$	$3 = \times 10^3$	$4 = \times 10^5$	$5 = \times 10^5$	$6 = \times 10^6$
$7 = \times 10^7$	$8 = \times 10^8$	$9 = \div 10$				

With the 3-symbol code, the first two digits indicate the base figures and the third the multiplier or number of zeros to add. Therefore 270 = 27R, or 27 pF, 331 = 330R or 330 pF, 472 = 4.7 K or 4.7 nF, 2R2 = 2.2R or 2.2 pF.

Aluminium electrolytic capacitors may use a 3 symbol code with numbers to indicate capacitance value in µF, plus a letter to indicate the voltage rating as follows: C = 6.3 V, D = 10 V, E = 16 V, F = 25 V, G = 40 V, H = 63 V.

The position of the letter in the code indicates the decimal point in the capacitance value. For example, F47 = 0.47 µF 25 V, 3E3 = 3.3 µF 16 V, 22C = 22 µF 6.3 V.

SMD soldering technology

Allied to the use of SMDs are the changes in soldering technology that have occurred over the recent past. What was at one time seen as an acquired practical skill is now seen as the application of a section of chemical science, a feature that has also made a significant contribution to the improvements in system performance.

The use of smaller circuit boards also leads to a further small improvement in RF performance. The printed circuit boards (PCBs) or substrates used with SMDs normally carry no through holes for conventional components; the component lead outs are soldered directly to pads or lands provided on the metallic print. This feature has led to the development of new soldering techniques which impart further advantages. In manufacture, the soldering methods that are used lead to improved connection reliability, which in turn leads to a reduction of costs. The particular technique employed in manufacture has a bearing on the way in which SMDs can be handled during servicing.

Wave soldering: The components are attached to the solder resist on each PCB, using an ultra-violet light or heat-curing adhesive. The boards are then passed, inverted, over a wave soldering bath with the adhesive holding the components in place, whilst each joint is soldered.

Reflow soldering: A solder paste or cream is applied to each pad on the circuit board through a silk screen, and components are accurately positioned and held in place by the viscosity of the cream. The boards are then passed through a melting furnace or over a hot plate, to reflow the solder and make each connection.

Vapour phase reflow soldering: This more controlled way of operating the reflow process uses the latent heat of vaporization to melt the solder cream. The boards to be soldered are immersed in an inert vapour from a saturated solution of boiling *fluorocarbon liquid,* used as the heating medium. Heat is distributed quickly and evenly as the vapour condenses on the cooler board and components. The fact that the soldering temperature cannot exceed the boiling point of the liquid (215°C) is an important safety factor.

Soldering and component heating

Electronic components experience distress at all elevated temperatures. For example, in wave soldering, most baths have an absolute limit of both temperature and time – normally, 260°C for no more than 4 seconds. As the damaging effects of heat are cumulative, manufacturing and servicing temperatures have to be kept to a minimum. The use of a low melting point solder is therefore crucial. The common 60/40 tin/lead solder (MP = 188°C) is not suitable for SMD use. The lowest melting point tin/lead alloy (eutectic alloy) solder has a melting point of 183°C, and this also is not suitable.

Components often have silver or gold-plated lead-outs to minimize contact resistance. Tin/lead solder alloys cause *silver leaching;* that is, over a period of time the solder absorbs silver from the component and eventually causes a high-resistance joint. This can be avoided by using a silver-loaded solder alloy of 62 per cent tin, 35.7 per cent lead, 2 per cent silver and 0.3 per cent antimony. Such a solder has a melting point of 179°C. Tin also tends to absorb gold with a similar effect, and this is aggravated by a higher soldering temperature. This latter alloy is thus particularly suitable.

The fluxes used as anti-oxidants are also important. An effective flux improves the solderability of the components, the rate of solder flow and hence the speed at which an effective joint can be made. The flux used for SMD circuits should either be a natural organic resin compound, or one of the newer equivalent synthetic chemical types.

Lead-free environment

During the next decade it is possible that a move will be made to introduce regulations that reduce the reliance on lead. Already research work is being carried out using a replacement solder such as tin/silver/copper alloys with a melting point that is about 40°C higher than the tin/lead product that has been in use for many years.

Question 8.3

Why should an SMC chip that has been removed, tested, and found to be good, not be replaced in circuit?

SMC values and tolerances

Resistor values and tolerances for SMD are the same as those used in more conventional applications. However, very high values (megohm ranges) are seldom employed with SM circuit boards. Because of the block-like structure and the good heat dissipating properties of the lead contacts, the power dissipation might be higher than expected. Typical common ratings range from $^1/_{10}$ to $^1/_4$ watts. Capacitors also have a normal range of values and tolerances except for the non-availability of very high capacitances. The common range of values extends from a few pF up to several µF but the working voltage ranges tend to be somewhat lower. A common range includes 20 nF for 50 volts up to 2 µF at 6.3 volts operation, showing that capacitances value tends to fall with a rise in working voltage.

Both resistors and capacitors are also manufactured in multiple units, typically up to 8 resistors or 4 capacitors in a single block.

Inductors are available ranging from 1 nH to 1 mH at up to 3 amps of saturation current. Similarly, transformers that cover frequencies ranging from audio applications through low RF at around 6 MHz are currently in use.

Device packaging

Having become familiar with single in line (SIL) and dual in line (DIL) devices, the surface mount technology now creates a plethora of new and sometimes obscure abbreviations. These include such as *small in-line packages* (SIP), SO for *small out-line devices*, (even SOL for *small out line*), SOT for *small out-line transistors* and more recently, SOC for *systems on a chip* where a few million transistors are fabricated into one IC.

Furthermore, complex ICs may be directly mounted on to a circuit board or plugged into a socket and this introduces a few more abbreviations. Large ICs with contacts arranged around 4 sides of a square are described as plastic quad flat packs (PQFP), or quad flat packs (QFP). Similar ICs may be mounted in a socket (J-lead contacts) described as a carrier (plastic leaded chip carrier or PLCC).

A further development involves the use of ICs with several rows of contacts arranged around the 4 undersides of the square chip package and these terminate in sometimes more than 400 contact bumps. Each of these are soldered directly to matching pads on the circuit board by one of the methods described above. This ball grid array (BGA) system creates real difficulties for repair or replacement. The only safe way to remove the chip is to use a hot air gun and then lift off the chip as quickly as possible. For replacement purposes, pads of pre-aligned solder bumps are available in a grid format to be used when replacing BGA devices.

Activity 8.2

It is very important to acquire a good degree of competence in handling SMDs, particularly the removal, testing and refitting of components. The removal can quite easily cause damage to the tracks on the circuit board which leads to further problems. Much useful experience can be gained from working on scrap boards where additional damage can only be good for experience.

Fault finding on SMC boards

As with discrete component circuit boards, the *divide and conquer* or *half split* technique has much to recommend itself in terms of time spent on fault finding with SMD boards. Locating the faulty area by signal tracing is applicable to both analogue and digital circuits. Once the faulty area has been located, the component or IC will have to be tested. L, C and R devices are usually fairly straightforward. Disconnect the suspect component and check on the circuit board for any unexpected short or open circuits associated with the contact pads that might have caused the fault. Test the displaced component to determine that it is in fact faulty before replacing with a new one. Even if doubt exists that the component might be serviceable, remember that the heating effects of removal and refitting could lead to an early failure. It is therefore often advantageous to fit a replacement.

To test a suspect IC, measure all the d.c. voltage supply points for correct value. Check for the same voltage levels on adjacent pins and check the continuity of earth lines. If signals are present at the input, then if the IC is serviceable and not overheating, then signals should be found at the outputs. However, there are a few exceptions to this rule where some power output stages, voltage stabilizer circuits, or similar may have shut down through an excessive rise in temperature.

If an IC has failed, it is important to ascertain why before fitting a new one. Check all the supply lines for open or short circuits whilst the IC is out of circuit. Check particularly for any adjacent pins that have acquired the same voltage levels. This might be caused by a short circuit somewhere in the circuit tracks.

Activity 8.3

Remove, test and if found to be serviceable, replace a number of resistors, capacitors and inductors on a SMD circuit board. Repeat the same exercise with transistors and ICs.

Activity 8.4

Estimate the power dissipated by a resistor on a SMD circuit board. Measure the value of the resistor with the power off and then volts drop across under power. Power dissipated is then V^2/R. What are the pitfalls of this method relative to measuring the volts drop and using the marked component value? In the first case the resistor may have a number of shunt components that give rise to a false resistance reading. In the second case, the resistor may have changed value through passing an abnormal current. This is probably one of the most difficult tasks because removing the component changes the circuit conditions and even an added meter has a similar effect.

Servicing SMD circuits

For the larger centralized service department, a soldering rework station might well be cost effective. This might include a small portable vapour phase soldering unit, such as the Multicore Vaporette, which is particularly suited for small batch work as commonly found in such establishments. For the smaller service department, much ingenuity might be needed when dealing with SMD circuits.

The method used to remove suspect components may vary with the number of leads per device. For a device with only two or three leads, a fine soldering iron, in conjunction with a *solder sucker* or *solder braid* can be successful. For multi-pin ICs, two methods are popular. One involves directly heating the soldered connections with a carefully temperature-controlled hot-air blast. The other method uses an electrically heated collet or special extension to a conventional soldering iron, to heat all pins simultaneously. When the solder flows, the component can be lifted away.

In a simpler way, all the leads can be severed using a pair of cutters with strong, fine points. The tabs can then be removed separately. This is not quite so disastrous as it appears. Any suspect component that is removed, and subsequently found not to be faulty, should *not* be reused; the additional two heating cycles are very likely to lead to premature failure. As the desoldering of even a few leads can be difficult, it is important that the circuit board should be firmly held in a suitable clamp.

Even after all the solder has been removed, a problem of removing the component may still exist if an adhesive was used in manufacture. Although care should be exercised when prising the component off the board, damage to the printed tracks is unlikely. The adhesive should have been applied only to the solder resist. After component removal, boards should be examined under a magnifier to check for damage to the print.

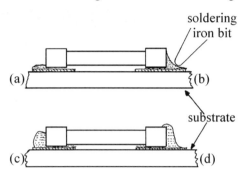

Figure 8.2   Soldering chip type SMDs

Before commencing to fit the new components, the solder pads should be lightly pre-tinned, using the appropriate solder and flux. Each component will need to be precisely positioned and firmly held in place whilst the first joints are made. For multi-leaded components, secure two diagonally opposite leads first. The soldering iron used should not exceed 40 watts rating and should not be applied to a joint for more than 3 seconds. The heat from the iron should be applied to the component via the molten solder and not direct as indicated, in Figure 8.2(b). The final joint should have a smooth 45° angled fillet. Figure 8.2 also shows three incorrectly made joints: (a) has used too little solder, and this produces a high-resistance connection; (c) has been made with the application of too little heat, which will also lead to a high-resistance connection; in (d), too much solder has been applied and this has probably resulted in excessive heating of the component.

Component reliability   Experience suggests that the most over-stressed discrete components in a system are the capacitors, particularly the electrolytics. If these are operated at too high a temperature, the electrolyte tends to dry out or leak. In the first case, this leads to a significant reduction in capacity with the attendant loss of performance. In the latter case, the leakage can create corrosion on the circuit board to generate open circuit tracks. When replacing a capacitor it is therefore important to ensure that not only is the capacitance value, voltage rating and ESR (equivalent series resistance – typically 2×10^{-2} Ω) are correct, but also that the temperature rating will not be exceeded. Typically for television and satellite receiver applications, a 105°C rating is specified but upgrading this to 125°C will often improve reliability. Again, if these devices are operated at too low a DC voltage, the dielectric material will start to decay. At first this causes the capacitance value to rise with the thinner dielectric, but this quickly punctures and becomes a short circuit. Modern circuit boards tend to be liberally populated with these devices and experience has shown that when one fails the others are quite likely to follow suit in quick succession.

Answers to questions

8.1   Early failure.

8.2   The LNB of a satellite receiver.

8.3   There is a considerable risk of damage because of re-soldering a component that has already been heated to remove it.

9 Safety testing and EMC

Portable appliance testing (PAT)

It is important to point out that there are generally two earth points within electrical equipment. The *primary earth ground* (PEG) or protective earth is designed to provide protection against electric shock by the user and the *secondary earth ground* (SEG) or signal earth is used to provide a return path for signals within the equipment. Great care needs to exercised where the two earths are connected together within a system. Each item of electrical equipment is connected to the mains supply via a green/yellow earth cable.

Portable appliances are described as those equipments that are powered via flexible leads plugged into a mains single phase electrical supply of 230 V a.c. Industrial equipment may be similarly connected to 3 phase, 400 V a.c. mains but are covered by a different class of regulation.

Within the single phase group there are three classes of equipments, defined as follows:

Class 1. With this equipment, the basic electrical insulation is additionally supported by providing a connection between the exposed metal parts and the protective earth provided by the mains power supply. This technique is employed in all industrial equipment.

Class 2. With this equipment, the protection does not rely on the basic insulation, but by additional internal insulation and the avoidance of any electrical connection to any exposed metal work. These are often described as being double insulated and are suitable only for domestic applications.

Class 3. These equipments are designed for use with *Safety extra low voltage* (SELV) supplies that operate at less than 25 V a.c. RMS or 60 V d.c. Rechargeable battery operated portable tools fall into this class.

All appliances connected to the mains supply via flexible leads and plug and socket connectors should be examined and tested at least annually by a competent person.

- Flexible leads should not be frayed, split, sharply kinked, cut, or too tightly clamped.
- Damaged leads should be replaced immediately.
- Cables must be securely fastened both at plug and the powered equipment.
- The supply cable to a heavy piece of equipment should be connected in such a way that the plug and socket will separate if the cable is pulled.
- The live end of the connector should have no exposed pins that may be touched.
- Every connection should be both mechanically and electrically sound.

- The cable must be well secured with no danger of working loose and with no stray strands of wire.
- Finally, be aware that hot soldering irons can easily produce damaged power leads.

It is important to be able to test and record the earth bond resistance, earth loop impedance and insulation resistance between line, neutral and earth after any repairs. This will ensure that the work carried out meets the standards required by law.

Question 9.1

Name the two earth connections commonly found in electronic equipment.

Earth loop impedance and continuity

A range of suitable test units is available from manufacturers such AVO, Megger Ltd, and Seaward Ltd which are readily portable and simple to use. The old fashioned Megger, which is still in existence, uses a hand cranked generator to provide the test voltages with the test results being indicated on a moving coil type meter. The modern versions are either driven from a mains power supply or a 9 volt battery to provide the necessary high test voltages. Most test sets have at least 2 ranges, 0 to 200M and infinity for insulation resistance and 0 to 2R for continuity. The indication of the results is displayed on an LCD screen which is supported by a battery backed semiconductor memory. The results can then be down-loaded into a PC memory for future reference.

The continuity and resistance of the earth lead should be tested whilst passing a current of 25 amps for five seconds which is high enough to open-circuit a partially fractured cable. The earth bond resistance should be in the order of 0R1 and not greater than 0R5 if the appliance plug is fused at 3 amps or less. This test is carried out by inserting the appliance plug into the test set and connecting the earth test lead to its conductive casing. With this test the series resistance of the mains cable needs to be taken into consideration but commonly this combined resistance will be less than 0R5.

The insulation resistance will require a 500 volts d.c. test voltage and for an earthed appliance the live and neutral pins are shorted together and the resistance between them and the earth connection is measured. The test voltage is applied for 5 seconds as the insulation resistance is being measured. Typically this should be at least 2M. For a double insulated device that carries the double square symbol, the test lead is connected to any exposed metal work and the insulation resistance should be higher than 7M.

It is also useful to attach a self adhesive label to the device recording a serial number, perhaps in bar code, referring to the test data. This will help to make workers in the field aware of the continuing need to maintain this standard. PAT testers are available that have the facility of storing the test data so that it may be downloaded into a PC data base for future record purposes.

On occasions when it is necessary to measure the state of the electrical distribution network in, say, a small workshop, the job is only slightly more complex but involves the following definitions.

R_1 = resistance of the phase conductor

R_2 = resistance of the protective earth conductor

Hence $R_1 + R_2$ = the loop resistance

Z_s = earth fault loop impedance

Z_e = the earthing impedance, external to the circuit

$Z_s = Z_e + R_1 + R_2$, and generally in such an application, any stray reactance effects can be ignored.

Then proceed as follows;

1. Disconnect the network from the supply by means of the main circuit breaker.
2. Strap the phase to earth at the distribution board.
3. Test the loop and insulation resistance between phase and earth at each outlet socket.
4. Record this information.

Activity 9.1

Take the appropriate safety precautions before starting this exercise. Ensure that the environment is dry and you are standing on a rubber mat and use one hand only. With a neon probe, identify the live line of a mains supply outlet socket. Then using an appropriate meter, measure and record the voltages between the live and neutral conductors and earth.

Take about 1 metre length of 3 core mains flexible cable and short the live conductor (brown) to the earth line (green/yellow) and measure and record the loop resistance at the open end. Now measure and record the insulation resistance between the shorted pair of conductors and the neutral wire (blue).

Mains polarity

The polarity of the mains supply (which is something of a misnomer because the voltage on the live line is bipolar) can be checked by using a neon lamp or probe. This device glows when in contact with a connection at a potential greater than about 100 volts d.c. or a.c. Using a voltmeter referenced to earth, the live line voltage should read 230 volts a.c. per phase. Theoretically since the neutral line is connected to earth at the nearest sub-station, this should read zero volts. However, small induced voltages of perhaps 10 volts a.c. may in practice be measured between neutral and earth.

Activity 9.2

Using a range of mains powered electronic test equipment, (CRO, signal generator, electronic voltmeter, etc.), disconnect the equipment from the mains supply. With a suitable instrument or PAT tester measure and record the earth loop impedance, the earth bond resistance and insulation resistance to earth.

Activity 9.3

With suitable test instruments and using a reel of flexible 3 core mains power cable, measure and record the loop resistance when the live and neutral lines are shorted at one end. Measure and record the insulation resistance between the shorted pair and the earth line. Cut off about 1 metre length and repeat the exercise on the shorter piece of cable. Explain how these changes of value arise.

EMC/EMI

Electromagnetic compatibility (EMC) is defined as the ability of a device, equipment or system to function satisfactorily in its electromagnetic environment without introducing intolerable electromagnetic interference (EMI) to any other system, while at the same time, its own performance will not be impaired by interference from other sources. Such interference occurs as noise within a system and since this is considered as a destroyer of information, it is often a limiting factor in a communications system.

Interference can be divided into two categories, natural and man-made. The former often results in electrostatic discharge (ESD), whilst the latter usually results in power line surges. Natural sources of interference include ionospheric storms and lightning. Man-made interference commonly results from high current switching operations and radio frequency generators. With the expansion in the use of portable digital communications systems such as computers and cordless mobile telephones, the sources of interference are growing rapidly.

A direct lightning strike is not necessary to cause havoc with a communications system. Even cloud to cloud discharges can easily set up high electric fields that can influence power lines and overhead telephone lines that tend to act as aerials. Instances have been recorded of induced voltage spikes in the order of about 2000 volts, but of very short duration being developed on the normal 230 volts a.c. supply mains. Even if exposures to such discharges do not prove to be immediately destructive, they can introduce a latency failure mechanism because these effects are cumulative.

The rubbing action between dissimilar materials has long been a recognized phenomenon now referred to as *triboelectricity*. A triboelectric series lists the materials ranging from air, hands, asbestos, at the most positive, through the metals, to silicon and Teflon as the most negative. The further apart in this table, then the higher will be the electric charge between the two materials when rubbed together. Even the action of a person walking across a carpet or even sitting in a chair can generate very high electrostatic charges, depending on the clothing worn and level of humidity. Due to the $Q = CV$ joules relationship between charge and voltage, the action of a sitting person lifting their feet off the ground reduces the body capacitance and this automatically causes an increase in the voltage level to increase the risk of ESD.

Counteracting the effects of EMC

Electromagnetic interference can enter communications equipment either by direct radiation or by conduction. The effects of the former can be reduced by screening or shielding, while reduction of the latter requires the use of suitable filters.

Connecting leads, particularly the longer ones make very good aerials for propagating interference problems. Such leads are also susceptible to breaking where they can vibrate. The loose ends then become the source of arcing or even short circuit problems.

Such leads should therefore be dressed close to the circuit boards or chassis and firmly clamped so that their capacitive pick-up effect is minimized.

Mains plug filters that have a low pass characteristic, with inbuilt shunts to discharge any excessive voltage spikes to earth, are very useful for restricting the ingress of mains borne interference, particularly in respect to digital and computing equipment. Protection for such equipment can avoid the *black-out* of services which lead to complete loss system functions. *Brown-out* is the jargon term used to describe partial loss of processed data.

- Holes in metal work should have dimensions that are less than about one tenth of a wavelength at the major operating frequency. A slot is a particular form of aerial that will radiate and accept signals very effectively at higher frequencies.

- All plug and socket type connectors should maintain good low resistance contacts, because high resistance generates heat and the resistance itself then creates a noise component.

- Shielding problems can arise after servicing by failing to replace all the screws and screen fixings correctly. Because of the wavelength effect, even the spaces between the screen securing screws can be critical in very high frequency equipment.

Figure 9.1 shows how a particular problem described as *earth loops* can arise.

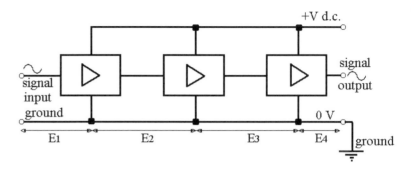

Figure 9.1   How earth loops develop

Because all three stages of the amplifier are earthed close to each stage, and the signal earth returns and ground earth are separated by conductor strips a number of voltages E_1, E_2, E_3 will be developed. These will range from d.c. values for the power supply to a.c. signals which are of varying phase. These will all add to make an earth loop that provides erroneous signals, instability or even distortion. The effect is even more pronounced at very high frequencies when the circuit operation can be improved by the addition of individual stage screening as shown in Figure 9.2. Feed-through capacitors that give a low pass characteristic and screened input and output connectors all help to isolate each stage from the others. The use of a common return to avoid the earth loops, together with overall screening further improves the system stability.

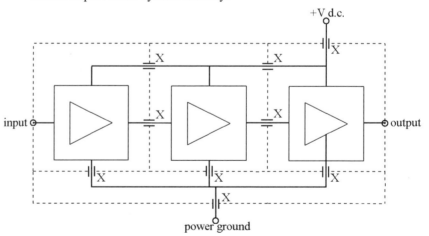

Figure 9.2   Improved circuit performance through the use of common earth points

Question 9.2

What does the abbreviation EMC mean?

CE Marking and Low voltage directive (LVD)

Both of these European Union (EU) directives are each very complex documents and their implications are closely allied in their relationship to the electronics and electrical manufacturing and servicing industries. Even the relative dimensions of the two circles that form the basis of the marker characters are specified. The CE marking signifies that the product or equipment to which it refers meets the *Health and Safety* features and complies with the appropriate regulations of LVD EN60950 and meets the emissions and immunity of EN50081/2.

- The standards have to be maintained throughout the product or equipments lifetime and additionally applies to second users and repaired items.

The LVD (low voltage directive) applies to items rated at between 50 volts and 1000 volts a.c., and 75 volts and 1500 volts d.c. It therefore applies to all domestic entertainment and most of the industrial equipment so powered.

The EN60950 directive which applies to IT systems and any equipment that may be connected to an IT system, is a lengthy document that chiefly relates to matters of safety, including flammability, for the users, operators and service personnel. In the interests of brevity, only the most important features of the directive can be included here. For specific information the reader is referred to the book list below.

All equipment is assumed to be in continuous operation unless otherwise stated on the rating label. Virtually all IT equipment is covered by this legislation, including Classes 1, 2, and 3 devices with pluggable connectors, plus fixed, portable and hand-held equipments.

The maximum earth leakage current is 3.5 mA which should be the trip setting of any associated earth leakage circuit breakers (ELCB). Even lithium batteries commonly used for memory back-up are specifically mentioned in the text, because these can be explosive if they are incorrectly replaced. During servicing, care must be exercised because equipment such as CRT type monitors, some laser type printers and photocopiers can make use of extra high voltages that exceed the LVD levels.

The LVD ratings are not intended for use in the following cases which are covered by other international specialist safety provisions;

1. In ships, aircraft, railways, explosive atmospheres, radiological and medical applications,
2. Domestic plugs and sockets and electricity supply meters,
3. Electric fence controllers and parts for goods and passenger lifts.

In addition, the directive does not apply to electrical equipment intended for export to third countries.

The CE marking is the only marking that may be used to certify that a manufactured product conforms to the standards incorporated in the relevant directives. It is a criminal offence not to mark a product that is covered by LVD or so mark a product that does not comply.

Essential points in the Directive

Equipment may be provided on the market only if, having been constructed in accordance with good engineering practice in the safety matters in force within the community; it does not endanger the safety of persons, domestic animals or property when properly installed, maintained and used in the applications for which it was intended.

The conformity to standards process
Third party test house route

Third party conformity assessment is normally carried out by a recognized test house. These organizations, who can prove their conformity with the EN 45000 series standards which covers the criteria for the operation and standards testing of new equipment are designated by the member states. These bodies may submit an accreditation certificate together with other documentary evidence.

Self declaration route

Under this route, the manufacturer takes full responsibility for the assessment, testing, documentation and declaration of conformity and the CE marking. A technical file (see below) or documentation must also be available on demand for the national enforcement authorities.

The technical file route

In order to certify conformity with the directives and the CE marking, the manufacturer must draw up a technical construction file covering the design, manufacture and operation of the equipment, prepare an EC declaration of conformity and affix the CE marking.

Control of substances harmful to health (COSHH)

An introduction to this topic was included in the Level 2 Edition of this series, to which the reader is referred for basic background information.

The human effects of such hazardous substances requires specialist treatment and it is therefore important that first aiders should be chemically aware of the associated problems. Under the regulations, all harmful substances should be stored in a cool dry place and used with safety concerns as the primary consideration. The person responsible for this store should also be a trained chemical first aider. The safety information provided by the manufacturer should be stored where it is readily accessible to the appropriate member of staff that may need it in a hurry.

Common solvents such as trichloroethane, trichloroethylene and carbon tetra chloride (CTC) that have in the past been used for cleaning and degreasing are now banned under EU law. Because all chlorinated solvents have narcotic and anaesthetic properties, any such solvents that are still in use should be treated very carefully. (Chloroform, another member of the family, is a well known anaesthetic.) High levels of exposure to this group of substances can lead to kidney and liver damage. Of the group, only trichloroethane is non-carcinogenic. Although these substances are non-flammable, when chlorinated fluids are exposed to heat, they liberate the chemical warfare gases known as phosgene and hydrogen chloride. The chlorinated solvents can enter the human body either by inhalation or by skin absorption.

Question 9.3

How does the CE marking system affect the service engineer?

The use of chlorofluorocarbons (CFC) are progressively being phased out largely because they are destroyers of the upper atmospheric ozone layer which protects against ultra-violet radiation from the sun. Apart from refrigeration applications CFCs have been used as degreasing agents and freezer sprays. The very low boiling point can create frostbite due to the rapid evaporation and can strip the fatty tissue from the surface of human skin. Therefore great care needs to taken when using up old stock of CFCs.

Hydroflourocarbons (HCFCs) that are being used as temporary replacements for CFCs are very good and non-destructive cleaners and non-flammable. However, under extreme heat they break down to generate phosgene, hydrochloride and hydrogen fluoride gases so need to be very carefully stored.

Ethyl alcohol or ethanol, together with its relative methylated spirits, are very good cleaning agents but are highly flammable, highly toxic and can cause liver damage if drunk. This group also absorbs water fairly easily and this slows down the evaporation process.

Isopropyl alcohol (IPA) which is highly flammable, is a relatively safe cleaning agent, although its near neighbour, N-propanol is classified as being toxic.

Petrol and benzene are both extremely good degreasing and cleaning agents but are highly flammable and explosive. They are harmful to health via skin absorption and the respiratory system and are also carcinogenic.

White spirit which is commonly associated with paints is highly flammable and produces narcotic fumes which are harmful if they enter the respiratory system.

The solvents such as xylene and toluene that are constituents of some switch cleaners and adhesives, produce euphoria (the glue sniffing syndrome) which lead to serious problems such as respiratory failure and death, damage to the central nervous system and liver damage through skin absorption. Acetone and amyl acetate are solvents that produce toxic fumes and are also a fire hazard.

Many of the adhesives and sealants used in the servicing environment are also likely to create health risks when used at elevated temperatures. Long periods of exposure to the fumes from epoxy resins (2-part mixes) can also be a source of liver problems. Whilst silver loaded epoxy glues have a low resistance that makes them useful for refixing metal shields within a piece of electronic equipment, the period of exposure can be reduced by raising the temperature to about 40° to 50° in order to cut the curing time in half.

Solder fluxes (rosin or colophony) produce fumes when heated that can cause industrial asthma. Since the low-fume alternatives are also suspect, it is most important that all soldering operations are as far as possible carried out under a fume extractor hood. Fume extractor kits that mount directly onto a soldering iron are available but tend to make the work more difficult. Smoke extractor hoods incorporate a charcoal filter that needs to be periodically replaced. The fumes from the lead in solder are also problematical, but whilst a smoke hood is helpful, the use of lead is included in regulations that are not within the scope of COSHH.

As with all accidents, events that are covered by COSHH must also be recorded in the workshop accident books which must be kept up to date by a responsible person and available for inspection by the appropriate authority.

NOTE

The European Union directives referred to in this Chapter cover several weighty volumes and are written in legalistic terms. The information presented above can only represent a brief resume of the salient points in each document. For further and more explicit explanations of the Directives, the reader can usefully refer to the following publications.

EMC for Product Designers, 3rd Edition, 2001, Tim Williams, Newnes, Butterworth-Heinemann, Oxford.

CE Conformity Marking, 2000, Ray Tricker, Butterworth-Heinemann, Oxford.

Practical Guide to the Low Voltage Directive, 2000, Gregg Kervill, Newnes, Butterworth-Heinemann, Oxford.

Communications Technology Handbook, 2nd Edition, 1997, Geoff Lewis, Focal Press, Butterworth-Heinemann, Oxford.

Answers to questions

9.1   PEG (primary earth ground) and SEG (secondary earth ground).

9.2   Electromagnetic compatibility, not causing excessive interference.

9.3   The equipment must, after servicing, still comply with the CE standards.

Unit 3

Outcomes

1. Demonstrate an understanding of d.c. power supplies to component level.

2. Demonstrate an understanding of low-frequency amplifiers, power amplifiers and operational amplifiers to component level.

3. Demonstrate an understanding of oscillators, multivibrators and waveform generator circuits to component level.

10　D.c. power supplies

Revise from Level-2 work, the use of diodes in half-wave, full-wave and bridge rectifier circuits with reservoir capacitors.

Activity 10.1

Connect the bridge rectifier circuit shown here, taking care that all the diodes (type 1N4001 or equivalent) are correctly connected. If one diode is incorrectly connected, it, and the diode which is in series with it across the a.c. input, will go open-circuit due to excess current.

Set the oscilloscope controls to 5 V/cm and 10 ms/cm, and switch the circuit on.

Examine the output waveform. Measure the d.c. voltage, using a multimeter on its 25 V range connected across the load resistor. Remove any one diode, and study the effect this has on the waveform.

Now connect into the circuit the reservoir capacitor C_1. Again examine the waveform across the load, and re-measure the d.c. voltage with the multimeter. Switch the oscilloscope input to a.c. and to a more sensitive range (say, to 0.5 V/cm) so that the remaining waveform (called the ripple) can be seen.

Finally, connect either a 100R, 1 W resistor in place of the load resistor or a 220R, 1 W resistor in parallel with the existing one, and observe the change in the ripple waveform voltage.

Try the effect of removing any one diode from the circuit to simulate the effect of an open-circuit diode. Repeat your measurements of d.c. voltage, both with and without a load, and of ripple voltage on load. Explain why measurements taken on-load are more useful.

In all rectifier circuits, each diode will be reverse-biased for half of the a.c. cycle and conducting for the other half. The amount of this reverse bias depends on the type of circuit used, and is greatest for a half-wave rectifier feeding into a reservoir capacitor.

Table 10.1 shows the operating results which may be expected from rectifiers of the three types described, half-wave, full-wave and bridge, given an a.c. input of E volts peak (0.7E volts r.m.s.) and smoothing by a reservoir capacitor of adequate size. Note the large difference between no-load output voltage and the fully-loaded voltage.

Note that in all rectifier circuits, reversing the connections of the diodes reverses also the polarity of the output voltage.

Table 10.1   Rectifier circuits compared for peak input E volts

Circuit	D.c. output (no load)	Reverse voltage, each diode	Ratio $I_{d.c.}/I_{a.c.}$	D.c. output(full load)
Half-wave	E	2E	0.43	0.32E
Full-wave	E	E	0.87	0.64E
Bridge	E	E	0.61	0.64E

Question 10.1

What are the three advantages that are achieved by adding a large reservoir capacitor to a simple rectifier circuit?

Zener diode regulation

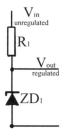

V_{in} unregulated

R_1

V_{out} regulated

ZD_1

Failure, rectifier circuits

A Zener diode is used reverse-biased so that the junction breaks down to permit current flow through the diode. This breakdown occurs at a precise voltage whose value depends on the construction of the diode, and causes no damage provided the current flow is not excessive. To prevent this, current must be regulated by connecting a resistor in series with the diode. The output of a Zener diode circuit is therefore a *regulated* voltage.

A simple regulator circuit is shown, left, with the output voltage across the diode being used to supply any other circuit or part of a circuit that requires a stable voltage. Many circuits, especially measuring and oscillator circuits, are adversely affected by voltage variation, which can be caused by changes in the supply voltage or by changes in the current drawn by the load.

Open circuit diode failures have the following effects:

• The d.c. output of rectifier circuits is either reduced or falls to zero. A half-wave circuit will have no output; full-wave and bridge circuits will have reduced output, with ripple at supply frequency if only one diode fails.

• A Zener diode regulator will give a higher voltage output, without regulation.

Short-circuit failures of diodes have the following effects:

• Rectifier circuits will blow fuses, and electrolytic capacitors may be damaged. A short-circuit diode will usually fuse itself, so becoming open-circuit. In a bridge circuit, a short circuit diode will usually cause another diode to become open-circuit.

• A Zener diode regulator will give zero output.

Regulation

The simplest possible power-supply circuits consist only of rectifiers and smoothing capacitors. In these circuits, the reservoir capacitor supplies the current to the load during the time when the diodes are cut off, and any ripple on the supply is reduced by filtering. Such circuits are adequate for many purposes, but they are too poorly regulated for use in circuits intended for measurement, computing, broadcasting or process control.

The *regulation* of a circuit is the term used to express the change of output voltage caused either by a change in the a.c. supply voltage or by a change in the output load current. A well regulated supply will have an output voltage whose value is almost constant; the output voltage of a poorly regulated supply will change considerably when either the a.c. input voltage or the output current changes. Table 10.1 showed how the voltage of a smoothed supply changes from no-load to maximum-load condition.

The change in output voltage caused by changes in the a.c. supply voltage is usually less great. A 10 per cent change in the a.c. supply will also change the output voltage of a simple power supply by about 10 per cent, the two percentage changes being almost identical.

Example: What will be the change in an unregulated 10 V supply when the a.c. supply voltage changes from 240 V to 220 V?

Solution: The a.c. voltage change is 20 V down on 240. The percentage change is therefore $20/240 \times 100 = 8.3$ per cent. Since 8.3 per cent of 10 V is 0.83 V, the 10 V supply will drop to $10.0 - 0.83$ V $= 9.17$ V.

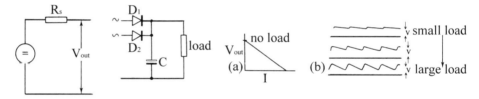

Figure 10.1   The causes of poor regulation with (a) V/I load graph and (b) ripple illustration

The effect of changes in the load current (which are themselves caused by changes in the load resistance) is more complicated, because there are two main causes (see Figure 10.1). The first is the resistances of the rectifiers, the transformer windings and any inductors which may be used in filter circuits, and by the internal resistance of the power supply. By Ohm's law, a change of current flow through such resistances must cause a drop in the output voltage.

The second effect is the voltage drop that takes place across the reservoir capacitor as more current is drawn from it. This voltage drop (V) can be quantified approximately by using the equation:

$$V = \frac{I \times t}{C}$$

where I is the load current in amps, t the time in seconds elapsing between one charge from the reservoir and the next, and C the capacitance in farads of the reservoir.

Example: By how much will the voltage across a 220 μF capacitor drop when 0.2 A is drawn from a full-wave rectifying circuit?

Solution: In a full-wave rectifying circuit, the time between peaks is 1 ms. Substituting the data in the equation $V = \dfrac{I \times t}{C}$ we get:

$$V = \frac{0.2 \times 0.01}{220 \times 10^{-6}} = 9\,V$$

A 50 V supply, for example, would give an output which drops to 41 V between peaks, and a 9 V peak-to-peak ripple at 100 Hz would be present.

Regulator circuits

A regulator circuit connected to a rectifier/reservoir unit ensures that the output voltage is steady for all designed values of load current or of a.c. supply voltage. Such a circuit can only do its job, however, if the rectifier/reservoir unit itself is capable of supplying the required output voltage (measured from the minimum of the ripple wave) under the worst possible conditions, when a.c. supply voltage is minimum and load current is maximum. The regulator will then prevent the output voltage from rising above this set value even when load current is small or the a.c. supply voltage goes high.

- A regulator requires a reference voltage that remains steady, and the action of the regulator is to compare the output voltage to this reference voltage and adjust the output accordingly. The reference voltage is obtained from a Zener diode.

Two basic types of regulator are used, a series circuit and a shunt circuit. The simplest example of the shunt type is a Zener diode regulator illustrated earlier. In this type of circuit, the amount of current drawn from the power supply is constant. When load current is maximum, the regulator circuit current is minimum. When the load takes its minimum current, the regulator circuit takes its maximum current. The arrangement is therefore such that load current plus regulator current is always a constant value. The constant current flowing through R produces a constant voltage drop across it. So if the input voltage remains unchanged, the output voltage will remain constant also.

The maximum dissipation of the Zener diode in this circuit is given by:

Zener voltage × maximum current

with dissipation in milliwatts if the current is measured in mA.

The maximum dissipation of the resistor is given by:

(Unregulated voltage – Zener voltage) × maximum current.

Example: A 5.6 V (5V6) Zener diode is used to supply a load which takes a maximum current of 15 mA. If the minimum desirable Zener current is 2 mA, and the unregulated voltage is 12 V, find (a) the value of series resistance which must be used, (b) the maximum Zener dissipation and (c) the maximum dissipation in the resistor.

Solution: With 15 mA flowing through the load and 2 mA through the Zener diode, total current is 17 mA. The voltage across the resistor is $12 - 5.6 = 6.4$ V, so that (a) the required resistance value, using Ohm's law, is $6.4/17 = 0.376$K or 376 ohms. In practice a 330 ohm resistor would be used, making the total current $6.4/.330$ mA $= 19.4$ mA. (b) At a current of 19.4 mA, the dissipation in the Zener diode is $5.6 \times 19.4 = 108.6$ mW; and (c) the dissipation in the resistor is $6.4 \times 19.4 = 124.2$ mW.

Series regulator

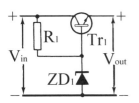

The other, and more important, way of regulating the output of a power supply is by means of a series regulator in which a transistor is connected between the supply and the load, as indicated here. This is a much more common type of regulator circuit.

When this type of regulator circuit is used, the transistor takes only as much current as the load, unlike the shunt regulator which takes its maximum current just as the load takes minimum current. The base voltage of Tr_1 is held constant by the regulator action of ZD_1 and R_1. The emitter voltage, and so the conduction, of Tr1 depends on the load voltage. If the demand for load current increases, the output voltage will tend to fall, increasing the forward bias of Tr_1. This allows it to pass more current to meet the demand. If the requirement for load current falls, this effect will be reversed.

Activity 10.2

Make up the simple series regulator circuit with the following component values:

$R_1 = 330R$ $ZD_1 = 5.6$ V $Tr_1 = 2N3055$ Vin $= 10$ V.

Connect this circuit to the power supply previously used, and draw a set of regulation curves for the stabilized circuit.

Measure also the ripple voltage at maximum load current (a) across the reservoir capacitor and (b) across the load.

More elaborate series regulator circuits use IC comparator amplifiers to drive the series transistor, and the whole circuit is in IC form. The use of IC regulators has now almost totally replaced discrete regulator circuits.

IC regulators

IC regulators are available in a bewildering variety of types, but the simplest are the familiar fixed voltage types such as the 7805. The final two digits of this type number indicate the stabilized output, 5 V for this example. The 7805 is a three-pin regulator which requires a minimum voltage input of 7.5 V to sustain regulation, with an absolute maximum input voltage of 35 V. The maximum load current is 1 A and the regulation against input changes is typically 3 to 7 mV for a variation of input between 7.1 V and 25 V. The regulation against load changes is of the order of 10 mV for a change between 5 mA and 1.5 A load current. The noise voltage in the band from 10 Hz to 100 kHz is 40 to 50 µV, and the ripple rejection is around 70 dB. Maximum junction temperature is 125°C, and the thermal resistance from junction to case is 5°C/W.

The 78xx type of regulator is used extensively for power supplies in both linear and digital equipment. A typical recommended circuit is shown in Figure 10.2 along with the diode bridge and reservoir capacitor. The capacitors that are shown connected each side of the IC are very important for suppressing oscillations and must not be omitted. In particular, the 330 nF capacitor at the input must be wired across the shortest possible path at the pins of the IC. The regulator should not be operated without a load, since this can cause the circuit to oscillate at a high frequency (around 5 MHz).

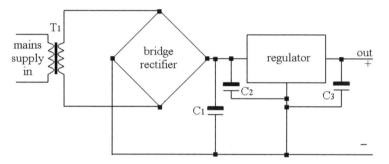

Figure 10.2   Using the 78xx type of regulator

The maximum allowable dissipation is 20 W, assuming an infinite heatsink, and the actual dissipation capabilities are determined by the amount of heatsinking that is used. If no heatsink is used, the thermal resistance of the 7805 is about 50°C/W, and for a maximum junction temperature of 150°C this gives an absolute limit of about 2.5 W, which would allow only 2.5 V across the IC at full rated current, an amount only just above the absolute minimum voltage drop.

The thermal resistance, junction to case, is 4°C/W, and for most purposes, the IC would be mounted on to a 4°C/W heatsink, making the total thermal resistance 8°C/W. This would permit a dissipation of about 15.6 W, which allows up to 15 V or so to be across the IC at the rated 1 A current. This is considerably more useful, since a 5 V supply will generally be provided from a 9 V transformer winding whose peak voltage is 12.6 V, making it impossible to cause over-dissipation at the rated 1 A current since the voltage output from the reservoir capacitor will be well below 12.6 V when 1 A is being drawn.

The 78 series of regulators are complemented by the 79 series, which are intended for stabilization of negative voltages. The circuits that can be used are identical apart from the polarity of diodes and electrolytic capacitors, and the range of currents is substantially the same as for the 78 series.

- If the regulated current that is required is greater than can be supplied from an IC regulator, then the main options are either to use the regulator to control a power transistor which is rated to pass the required current, or to use a switched mode power supply (SMPS, see later) rather than an IC regulator.

Protective circuitry

In the early days of IC regulators, the failure rate of regulator chips was very high, but later versions have added protective circuitry which has almost eliminated failures of the type that caused so many problems at one time. The main protective measures are for thermal protection and foldback overload protection, and the effect in each case is that the output from the regulator will drop to a level that reduces risk to the regulator or to the equipment fed by it.

Dropout

The *dropout* of a regulator is the minimum voltage difference which must exist between input and output in order to sustain the action of the regulator. The dropout for the 78 series of regulator ICs is at least 2 V and for some purposes, particularly for stabilizing supplies that are based on secondary batteries, this differential is too large. Low dropout regulators allow much lower levels of input voltage, typically down to 5.79 V for a 5 V output regulator.

Low-dropout regulators were originally developed for the car industry to provide stabilized outputs for microprocessor circuitry, and they would not be used for general-purpose stabilization for mains-powered supplies. Most low-dropout regulators feature additional protection against supply reversal, the effects of using jumper leads between batteries, and large voltage transients. Several types also feature inhibit pins, which allow the regulator to be switched back on again after it has been switched off by an overload.

Activity 10.3

Construct an IC regulator circuit, as illustrated in Figure 10.2, using the 7805. Measure the output for input voltage levels that range from 6 V to 12 V, and with load from 10 ohms to 1K.

Question 10.2

Comparing shunt and series regulator circuits, in which type would you expect the current in the regulator to be the same as the current through the load?

Failure of regulators

Failure in discrete regulator circuits is usually caused by excessive dissipation in the main transistor, shunt or series. In the series circuit the over-dissipation will have been caused by an excessive load current, unless the regulator is protected against short circuits.

- An o/c Zener diode will cause a shunt regulator to cease conducting. Its effect on the series circuit is the opposite, in that the output voltage will rise to the level of the unregulated supply.

- An s/c Zener diode will cause the shunt circuit to pass excessive current. It will cause the series circuit to cut off.

- Note carefully that all power transistors used in regulator circuits must be bolted to heat sinks of adequate size. Failure of IC regulators is usually due to thermal overload of regulators that do not have protection circuitry.

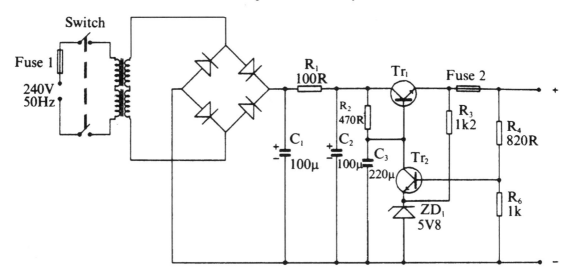

Figure 10.3 A rectifier-regulator circuit using discrete transistors so that the effect of faults can be demonstrated

Activity 10.4

Use the rectifier-regulator circuit of Figure 10.3 to assess the effect of faults, using either faulty components or simulating faults. For the circuit as shown, measure and tabulate the following:

(a) current flowing with no additional load

(b) d.c. output voltage

(c)   d.c. output voltage when a 1K load is connected
(d)   current flowing when the 1K load is added.

Now note the voltage levels at each terminal of each transistor, six voltage readings in all. Repeat these measurements when the following faults (one fault at a time) are introduced.

Tr_1 b-e o/c

ZD_1 o/c

R_3 equal to 22K

Tr_2 b-e o/c.

Explain the effect each fault has on the voltage levels.

Voltage multipliers

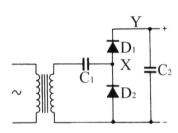

There are a few circuits that call for a high-voltage, low-current supply in which poor regulation is acceptable. Rather than wind a transformer especially to provide the high voltage, a voltage multiplier circuit (of which the voltage-doubler is the simplest example) is often used instead. On the negative-going half of the voltage cycle shown, left, C_1 is charged by current through D_1, so that Point X is at a d.c. voltage equal to the peak voltage of the a.c. wave.

At the peak of the positive-going half-cycle, the peak inverse voltage across D_1 is equal to twice the peak voltage (the previous peak charge, plus the peak a.c. value). This causes C_2 to charge through D_2 to the same level.

At line (supply) frequencies, the capacitor C_1 must be of large value, and must, of course, be rated at the full d.c. voltage. At higher frequencies, smaller values of capacitance can be used.

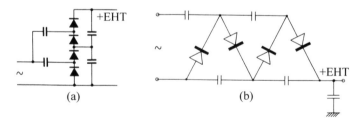

Figure 10.4   Voltage quadrupler circuits

Multiplier circuits of this type are commonly used to supply the high voltages required by the colour tubes in colour TV receivers. The circuit shown in Figure 10.4(a) is effectively two voltage-doublers connected in series, while Figure 10.4(b) shows the Cockcroft-Walton multiplier, named after its inventors. Circuits such as these are commonly used to obtain voltages as high as 25 kV for the final anode electrode of the tube.

PSU components

The components used in power supplies must be adequately rated for the voltage and current levels that will be used. Electrolytic capacitors can be used for low-voltage smoothing applications, but they should not be used for high voltage supplies, particularly above 500 V. For such applications, plastic dielectric capacitors must be used and for some specialized applications, oil-filled paper capacitors. Resistors are subject to a maximum voltage rating, often 350 V or less, and where higher voltages are used, chains of resistors connected in series must be used to keep the voltage drop across each individual resistor within the limit.

Power supplies that operate at low voltages usually deliver large currents, and computer power supplies in particular will supply 20 A or more at +5 V. Electrolytic capacitors used for smoothing such supplies (except for *switch-mode* supplies, see later) must be rated to have low equivalent series resistance and be able to withstand high ripple currents. Some supplies make use of a series resistor for monitoring current, and this resistor will have a very low resistance value and cannot be replaced with a conventional resistor.

> • The use of switch mode circuits allows plastic dielectric capacitors to be used in smoothing circuits rather than electrolytics, because the ripple frequency is much higher.

Thyristor power supplies

Power supplies that are designed to supply d.c. or a.c. at high power or high voltage need to be of rather different design. Many of them use thyristors as the controlling semiconductors (see Chapter 3).

A thyristor can be triggered on but not off. In the design of a power supply stabilized by thyristors, current shut-off is achieved by feeding the thyristor or thyristors either with a.c. or with unidirectional pulses that have a zero or a negative voltage peak. If a d.c. output is required, a diode must be placed in circuit between the thyristor and the smoothing capacitor to enable the thyristor to be switched off.

Two different types of control can be achieved with the aid of thyristors. First, in a *phase control* system, the level of average current flow is controlled by switching on the thyristor at different parts of the positive cycle, as illustrated. In such a system, the thyristor conducts on each positive cycle, but the phase of the gate signal varies according to the output required. A disadvantage is the large current surge which arises when the thyristor is switched on, causing radio-frequency interference (RFI). Phase control systems therefore require chokes and suppression capacitors to reduce the radiation of interference and also to prevent interference pulses from other thyristor-controlled equipment from affecting the thyristor gate.

The second type of control system, known as *zero-voltage switching* or *burst firing* system, can be used only when the load is a heater or a motor with a large flywheel. In a zero-voltage switching system (see waveforms, left), the thyristor is switched on only when the voltage between its anode and cathode is zero, so very little interference can be generated. A few complete cycles of the waveform are allowed to pass, then the gating pulses are removed, also at the zero voltage point. The repetitive waveform to the load thus consists of a few complete cycles, followed by zero input. This method of control can therefore be used with loads which have a long

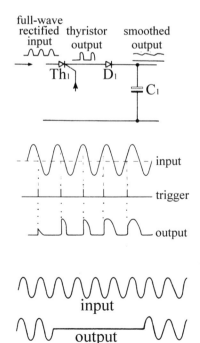

time-constant. Figure 10.5 shows a circuit in which a thyristor is used in the phase-control mode of operation to control average current flow through a load. On each half cycle, the thyristor remains non-conducting until C_1 has charged sufficiently to fire the unijunction. The current pulse through R_2 then fires the thyristor, and current passes through the load for the remainder of the half-cycle. If the load is connected in the line leading to the bridge rectifier, an a.c. load can also be controlled by this technique.

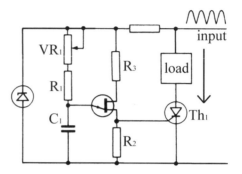

Figure 10.5   A typical thyristor power controller

Figure 10.6 shows part of a circuit which is used for controlling the speed of a shunt-wound d.c. motor. The diode D_1 is known as a *flywheel* (or *freewheel*) diode. Its function is to permit current to continue flowing freely in the circuit when the thyristor is not conducting.

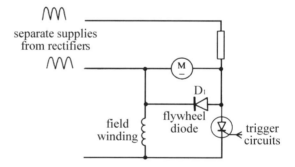

Figure 10.6   Part of a control circuit for a shunt-wound electric motor

Question 10.3

What is the advantage of using a rectifier circuit in which diodes have been replaced by thyristors?

Switch mode supplies (SMPS)

The most common configuration of linear-regulated power supply consists of a mains frequency transformer and rectifier, together with an IC series regulator. This latter simply behaves as a controlled series resistance to stabilize the output voltage. Such systems suffer from several serious disadvantages:

- They are most inefficient. It is unusual to find that more than 35 per cent of the input energy reaches the load. The remainder is dissipated as heat. The inefficiency is greater for low-voltage high-current supplies.

- The mains transformer is invariably large. Its size tends to be inversely proportional to the operating frequency.

- The reservoir and smoothing capacitors need to be large to keep the ripple amplitude within acceptable bounds. This is particularly difficult for low-voltage supplies.

- Because the series transistor (or transistors) is operated in the linear mode they must be mounted on large heat sinks.

If the operating frequency can be increased significantly, both the transformer and the filter capacitors can be reduced in size. If the series transistor can be operated either cut-off or saturated, its dissipation will be greatly reduced. The power supply can then be made more efficient. Such operation can be achieved using a switched mode power supply (SMPS). These circuits can operate with efficiencies as high as 85 per cent.

The basic switching principle of the most common type of SMPS (sometimes called a Buck converter) is shown here. When the switch is closed, current flows through the inductor or choke L to power the load and charge the capacitor C. When the switch is opened, the magnetic field that has been built up around L now collapses and induces an e.m.f. into itself to keep the current flowing, but now through the flywheel or freewheel diode, D. The voltage across C now starts to fall as the load continues to draw current. If the switch is closed again the capacitor recharges. This switching cycle produces a high-frequency supply voltage.

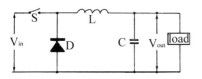

The duty cycle or switching sequence is shown in Figure 10.7 together with the output voltage V_{out}, that it produces. Increasing the on-period will increase V_{out} whose average level is given by $V_{in} \times t_{on}/T$. V_{in} can be regulated by varying the mark to space ratio of the switching period. Any unwanted ripple can be filtered off in the usual way.

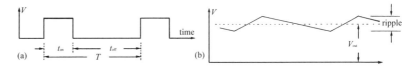

Figure 10.7   Switching action and waveform

The typical SMPS whose block diagram is shown in Figure 10.8, consists of a mains rectifier with simple smoothing whose d.c. output is *chopped* or switched at a high frequency, using a transistor as the switch. For TV applications this switch is commonly driven at the line frequency of 15.625 kHz. The circuit generally needs some startup arrangement that will ensure drive to the PWM switch when no d.c. output exists.

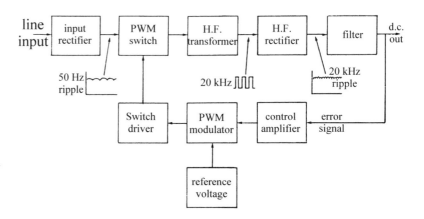

Figure 10.8   Block diagram of switched mode power supply

For industrial applications or computer power supplies the switching frequency is usually in the order of 20 to 25 kHz. The chopped waveform is applied to the primary circuit of a high-frequency transformer that uses a ferrite core for high efficiency. The signal voltage at the secondary is rectified and filtered to give the required d.c. output. This output is sensed by a control section that compares it with a reference voltage to produce a correction signal that is used in turn to change the mark-to-space ratio of the switching circuit to compensate for any variation of output voltage. This action is effectively pulse-width-modulation.

The ripple frequency of 50 Hz at the input has been changed to a frequency of 20 kHz at the output so that the smoothing and filter capacitors can be reduced in value by the ratio 20,000:50, equal to 400 times. No electrolytic capacitors need be used.

The oscillator/rectifier part of the circuit can be operated from a battery or any other d.c. input, so that it becomes a device for converting d.c. from a high voltage to one at a lower level. Another option is to use a transformer whose input is the chopped high-frequency voltage, with several outputs that are rectified to produce d.c. at different voltage levels. Only one of these levels can be sampled to provide control, so that only one output is stabilized against load fluctuations, though all are stabilized against input fluctuations.

SMPS circuits are universal in small computers, because of the need to regulate a low voltage supply at a high current. The usual circuitry rectifies the mains voltage directly (using no input transformer) so that the early stages operate at high voltage and low current, and a conventionally regulated supply is used to operate the control stages, ensuring that these are working at startup. A transformer for the high-frequency voltage provides for isolation from the mains and for voltage output of +5 V (main output at high current) along with –5 V, +12 V and –12 V. A complete SMPS circuit can be obtained in IC form, and for higher outputs an IC can be used to control a high-power switching transistor.

The SMPS generates more radiated and line conducted noise than a linear supply. This can be reduced to acceptable levels by using:

- Mains input filters balanced to earth to give rejection of the switching frequency.

- Suitable design of output filter.

- Electrostatic screen between primary and secondary of the mains transformer.

- Efficient screening of the complete unit.

Question 10.4

What is the main advantage of using a high switching rate, such as 40 kHz, in a switch-mode supply? What is the main disadvantage?

Excess-voltage trip

Figure 10.9 shows the layout of a typical excess voltage protection circuit using the *crowbar* principle. Tr_1 is the switcher transistor which is part of the SMPS unit, so that the base connections are not shown here. D_2 is typically a 72 V Zener diode and Th_1 is a thyristor. When the output voltage is held at its normal level of, say, 65 V, D_2 is reverse-biased and non-conducting. There is, therefore, no voltage developed across R_1, and the gate voltage of Th1 is zero.

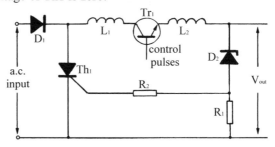

Figure 10.9   Excess voltage trip circuit for SMPS unit

If the output voltage rises above 72 V, D_2 conducts, draws current through R_1 and the gate voltage of Th_1 rises to trigger it into conduction. This applies a short circuit across the supply input, which can either blow the mains input fuse or trigger a thermal cut-out to break the circuit. The effect of causing the conduction of Th_1 is therefore like placing a very low resistance (the *crowbar*) across the supply rails when an over-voltage condition occurs.

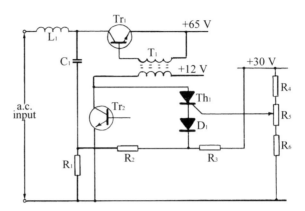

Figure 10.10   Excess current protection circuit

Excess current trip

The circuit shown in Figure 10.10 provides the SMPS unit with protection against excess current demands. Once again, Tr_1 and Tr_2 are transistors that are part of the conventional SMPS circuit, so that the base circuitry is not relevant here. L_1 offers a high impedance to the switching frequency current and is therefore by-passed by C_1 and R_1. This causes an a.c. voltage across R_1, which could cause Th_1 to conduct if the negative peak exceeds the gate voltage as set by R_5. If, under fault conditions, the current drawn from the 65 V line rises, there will be a corresponding increase in the a.c. voltage across R_1. On the negative excursion, Th_1 will therefore be fired to short out the switching pulses and cause the 65-volt line voltage to fall so that Th_1 again becomes a high resistance and the circuit resets. If the fault condition still persists, the process of trip and reset repeats at the switching frequency allowing only a limited current to be drawn.

Using a dummy load

Faultfinding and testing on any SMPS circuit can be difficult because of the way that the units are connected to each other and depend on each other. This is particularly so when the SMPS is part of a TV receiver for which the driving pulses for the SMPS are taken from the TV circuits. One important rule is that you should not attempt to operate the SMPS without a load.

If you need to isolate the SMPS for testing, you should connect a dummy load, and the ideal type of dummy load is an incandescent lamp with the appropriate power and current rating. Another safeguard is to obtain the input from a Variac™ autotransformer, allowing the a.c. input to be increased gradually. Yet another way of protecting the SMPS while under test is to include a 60 W lamp in series with the mains supply. The positive temperature coefficient of the lamp will guard against surges and protect against short-circuits.

Activity 10.5

The behaviour of SMPS can be effectively studied using switched-mode power supply kits. These not only provide valuable construction exercises and are useful as TTL type power supplies, but they can also be used to study the regulation and ripple characteristics of SMPS. In the absence of a suitable training kit, an ideal substitute is a PSU from a computer. These are now inexpensive and readily obtainable items that can provide considerable experience with SMPS, though it is not always easy to obtain a circuit diagram.

You should also record carefully voltages and waveforms for the SMPS working normally, and also in various fault conditions.

Inverters and converters

Although inverters are designed to convert d.c. power into a.c. power, they have some features in common with *converters* that are used to transform d.c. energy from a low-voltage level into d.c. at a higher voltage. The block diagram of Figure 10.11 shows some of these common features. An oscillator, usually running at an ultrasonic frequency (typically around 20 kHz) is powered from a d.c. source. This oscillator drives a transformer to provide either an a.c. output in the case of an inverter, or to power a rectifier/smoothing circuit for converter applications. It is important to note that in each case the total output power must be lower than that supplied from the d.c. source due to the energy losses in the transformer, etc.

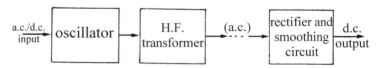

Figure 10.11 Block diagram of inverter/converter unit

Line filters and interference reduction

When using mains-powered equipment it is important that the minimum of noise and interference generated should be fed back into the mains. This is particularly important where data-processing equipment is operating in close proximity to high current industrial plant. Care is also needed where linear circuits are operating at low signal levels. Power supplies of the SMPS or oscillator types in particular should have their inputs well filtered to prevent pollution of the mains supply. For some linear circuits, the frequency of operation of the SMPS may have to be carefully chosen to avoid interference.

Interference can be continuous (as for SMPS) or transient (as when switching a supply on or off), and both radio-frequency and transient suppression will be needed for most types of power supplies.

$C_1=C_2=0.05\ \mu F$
$C_3=0.1\ \mu F$

$L_1=L_2=100\ \mu H$
$C_1=C_2=C_3=C_4=470\ pF$

Figure 10.12   Line interference filters

Typical filter circuits and component values are shown in Figure 10.12 (a) and (b). Inductors and capacitors used in these applications should have adequate working voltage and/or current ratings under fault conditions. The characteristics of such filters are low pass, with zero attenuation at 50 Hz and at least 30 dB over the frequency range 150 kHz to 50 MHz. Data processing equipment is particularly susceptible to mains-borne interference. Data may be corrupted by the high voltage transients induced from inductive loads on the mains supplies. Fortunately this problem is fairly easily solved by wiring devices called *varistors* across each pair of the mains supply wiring.

Varistors are particularly useful for transient suppression. These components are usually manufactured using zinc oxide or silicon carbide, with the former being often preferred because of its faster response. The characteristic of a varistor is shown here. The device has a high resistance below some critical voltage, but above this, it rapidly conducts to short circuit any large over-voltage condition. The working voltages of these devices, which can be used for both a.c. and d.c., range from about 60 V up to about 650 V.

Mains power outlets and mains lead plugs are now available for low power (less than about 2 kW) applications. These contain both filters and surge suppressors. They are particularly suited to the mains supplies for mini and microcomputer installations.

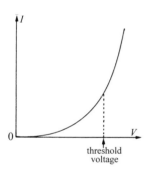

threshold voltage

Safety-critical components

Power supply units and other circuits that operate at high dissipation will usually contain safety-critical components. These are typically resistors with specified dissipation characteristics or capacitors that are constructed to avoid dangerous breakdown, and such critical components are usually marked as such on circuit diagrams.

The important point about such components is that they must be replaced, if found faulty, with the identical type of part, never with an ordinary component that seems to offer similar characteristics. Since a breakdown of such a component compromises safe use of the equipment, anyone replacing such a component with an ordinary one could be held responsible for the consequences of equipment breakdown.

Answers to questions

10.1 The output voltage is higher, the level more stable, and the ripple voltage is greatly reduced.

10.2 The series type.

10.3 The output level can be controlled by gating the thyristors at different points in the cycle.

10.4 The main advantage is smoothing without using electrolytics, the main disadvantage is the level of RFI.

11

Basic parameters

Amplifiers

The bipolar junction transistor (BJT) and the FET both control the current flow at their output terminals (collector and drain respectively); and in both cases this current at the output can be controlled by the voltage at the input.

The ratio:

$$\frac{\text{Change of current at output}}{\text{Change of voltage at input}} \text{ (with a constant supply voltage)}$$

is called the *mutual conductance* of the particular bipolar transistor or FET to which it applies. Its symbol is g_m. The values of mutual conductance obtainable from bipolar transistors are much greater than are those from FETs. In the mutual conductance graph shown in Figure 11.1, for instance, a 30 mV input wave gives a 1 mA current flow at the output. The mutual conductance, g_m, is therefore $1/0.03 = 33.3$ mA/V. MOSFET circuits are used for small-signal amplifiers mainly in IC form, though power MOSFETs are found in hi-fi amplifiers. The examples here are mainly for BJT circuits.

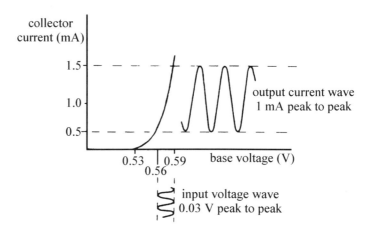

Figure 11.1 Mutual conductance graph

- The value of g_m for a BJT depends mainly on the steady value of collector current, Ic, and is approximately $40 \times Ic$ mA/V for any silicon transistor. If the collector load is Rc, then the gain is $40 \times Ic \times Rc$. Gain can be maximized by raising either steady collector current or the collector load resistance, or both.

The method of operation for signal amplification is as follows. A signal voltage, alternating from one peak of voltage to the other, at the input produces a signal current, alternating from one peak of current to the other, at the output. To convert this signal current into a signal voltage again, a load is connected between the output terminal and the supply voltage. In a d.c. or an audio amplifier, a resistor can be used as the load; but for IF or RF amplifier circuits a tuned circuit (which behaves like a resistor at its tuned frequency) is used instead.

When the connection is in the common emitter mode, the use of a load resistor causes the output voltage to be inverted compared to the input voltage. A *higher* steady voltage at the input causes a greater current flow at the output. A larger voltage is therefore dropped across the load resistor, which causes the output steady voltage to be *lower*.

Any amplifying stage can give current gain, voltage gain or power gain, and the amounts of gain which can be obtained depending on the way the circuits are connected. Figure 11.2 shows the three possible amplifying connections of a single transistor, with their relative gain values (bias components not shown). See also Chapter 3 – the circuits and their properties are restated here in the interests of simplicity.

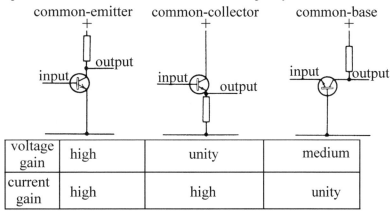

	common-emitter	common-collector	common-base
voltage gain	high	unity	medium
current gain	high	high	unity

Figure 11.2   Amplifying connections of a single transistor

Note that circuits using active devices such as bipolar transistors or FETs give power gain. The amount of this gain is calculated by multiplying voltage gain by current gain. Passive devices such as transformers can provide voltage gain or current, but not power gain.

The transfer characteristic

The behaviour of an amplifier can be clearly read from a graph of its output voltage or current plotted against its input voltage or current (for given values of load resistance and supply voltage). For example, the transfer characteristic of a small bipolar transistor is shown in Figure 11.3. An input current wave of 50 µA peak-to-peak produces an output current wave of 4 mA peak-to-peak. The current gain h_{fe} of the transistor under these conditions is thus:

$$\frac{\text{Collector current swing}}{\text{Base current swing}} = \frac{4\,\text{mA}}{50\,\mu\text{A}} = 8 \text{ (Note that 4 mA} = 4000\,\mu\text{A)}$$

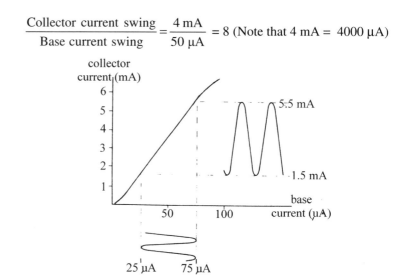

Figure 11.3   Transfer characteristic of a small bipolar transistor

The current wave that is produced is in this case an exact copy of the input wave, because the part of the graph which represents the values of input and output current is a straight line. Because a straight-line characteristic produces a perfect copy, such an amplifier is called a *linear* amplifier. If the input signal had been greater, with peak values 0 and 100 µA, the output signal would not have been a perfect copy of the input wave because these values of input would make use of the part of the characteristic that is not a section of a straight line.

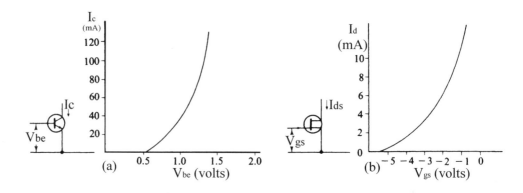

Figure 11.4   The mutual characteristics of (a) a typical bipolar transistor and (b) a typical junction FET

Figure 11.4 shows the plots of the output current/input voltage characteristics of (a) a bipolar transistor and (b) a junction FET. The curved shape of these characteristics shows that reasonably good linear

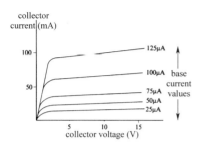

Bias

amplification is possible only if a small part of the characteristic is selected for use. It is not, for example, possible to use an input voltage of less than 0.6 V for the bipolar transistor.

Another way of looking at transistor or FET action is by drawing the *output characteristic*. In the illustration, left, each line is the graph of *Ic* plotted against V_c for a given base current I_b. The difference in spacing shows that amplification will be non-linear. In other words, if the output characteristic lines are drawn for equal changes of input voltage or current, the unequal spacing of the lines indicates that the transfer characteristic must be curved, producing non-linear effects.

The mutual characteristics of the bipolar transistor shown in Figure 11.4(a) make it clear that if such a transistor is to be used as a linear amplifier, the output current must never cut off nor must the output voltage be allowed to reach zero (its *bottomed* condition).

Ideally, when no signal input is applied, output current should be exactly half-way between these two conditions. This can be achieved by biasing, by supplying a steady d.c. input which ensures a correct level of current flow at the output. A correctly-biased amplifier will always deliver a larger undistorted signal than will an incorrectly-biased one.

A correctly-biased amplifier operating in such conditions, with the output current flowing for the whole of the input cycle, is said to be operating under Class A conditions.

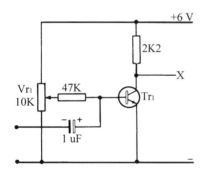

Activity 11.1

Connect up the circuit shown here, in which Tr1 can be any medium current silicon NPN transistor such as 2N3053, 2N1711, 2N2219 or BFY60. Set the signal generator to deliver a signal of 50 mV peak-to-peak at 1 kHz when connected to the amplifier input. Connect the oscilloscope to the output terminals, with the Y-input of the oscilloscope to Point X. Set the oscilloscope Y-input to 1 V/cm and the timebase control to 1 ms/cm, and switch on the oscilloscope. When the trace is visible, adjust the potentiometer Vr_1, to its minimum voltage position, and switch on the amplifier circuit.

Note that there is no output from the amplifier because it is incorrectly biased. Gradually increase the bias voltage by adjusting Vr_1 until a waveform trace appears. Draw the wave-shape.

Continue to adjust the bias voltage until the waveform seen on the oscilloscope screen is a pure sine-wave it may be necessary to adjust the amplitude of the input wave to achieve this.

Then increase the bias still further until distortion becomes noticeable again, and sketch this waveform also. You will see that with too little bias the amplifier cuts off, causing the top of the waveform to flatten.

With too much bias, the amplifier bottoms, causing the bottom of the waveform to flatten.

Bias circuits

Three types of bias circuit are illustrated in Figure 11.5. The simplest uses a single resistor connected between the supply voltage and the base of the transistor, Figure 11.5(a). This type of bias is not used for linear amplifier stages because it is difficult to find a suitable value of bias resistor. In this simple type of system, the value of resistor for correct biasing depends on the value of current gain (h_{fe}) of the transistor, so that a bias resistor suitable for one transistor will not work properly with another even if it is of the same type number. The value of bias resistor will always be critical, so that one preferred value of resistance is too low and the next in the series too high.

In addition, changes in the transistor characteristics as its temperature alters will cause the bias setting to change. This is because the voltage needed between the base and the emitter for a given collector current decreases as the temperature of a silicon transistor rises. With the simple system of bias, this change in forward voltage causes more base current to flow, and so more collector current flows as the temperature increases. Unless collector current is limited by a load resistor, the additional current will heat the transistor, so causing current flow to increase still further until the transistor is destroyed.

This process, called *thermal runaway*, is much less common nowadays, when use of silicon transistors is so general, than it was when germanium transistors were used extensively.

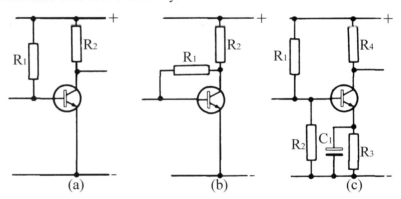

Figure 11.5   Bias systems (a) simple, (b) current feedback, (c) fixed base voltage type

The circuit shown in Figure 11.5(b) represents a considerable improvement, because the bias resistor is returned to the collector of the transistor rather than to the fixed-voltage supply line. This small change makes the bias self-adjusting, and the bias is now said to be stabilized.

The connection of the bias resistor as shown causes d.c. feedback, which means that the level of d.c. voltage at the collector affects the amount of d.c. bias current at the base.

See what happens in two opposite cases. First, a change either in the transistor itself or in the load which causes collector current to increase will, because of the presence of the load resistor, cause collector voltage to drop. By Ohm's Law, this will reduce current flow through the base resistor, so reducing base current, and so reducing collector current back to near its original value. Alternatively, a change that causes collector current to drop will make collector voltage rise, so passing more current through the bias resistor, causing more base current to flow, thereby increasing collector current back to nearly its original value.

All negative-feedback systems act in a similar way, tending to keep conditions in an amplifier unchanged despite other variations. A.c. feedback and its effects will be considered later on.

The third bias circuit, shown in Figure 11.5(c), is the most commonly-used of all for discrete transistor amplifier circuits, either BJT or FET. The negative feedback system of biasing is the main (and usually the only) method of biasing IC amplifiers, see later in this chapter. A pair of resistors is connected as a potential divider to set the voltage at the base terminal, and a resistor placed in series with the emitter controls the emitter current flow by d.c. negative feedback. Note the emitter current is practically equal to the collector current flow ($I_c = I_e + I_b$, but I_b is very small).

In this type of circuit, the replacement of one transistor by another has little effect on the level of steady bias voltage at the collector, and changes caused by altering temperature similarly have very little effect. This biasing arrangement is therefore ideal for use in mass-produced circuits which must behave correctly even when fitted with transistors having a wide range of values of h_{fe}.

Activity 11.2

Set up the circuit shown in Figure 11.6(a), preferably on a solderless breadboard such as the RS components range of prototyping boards.

Switch on and adjust the potentiometer until collector voltage is exactly half the supply voltage. Note this value of resistance.

Now connect a 470K resistor in parallel with the bias resistor. Note the new value of collector voltage. Remove the 470K resistor, and connect a 6K8 resistor in parallel with the load resistor. Note the value of collector voltage now. Remove the 6K8 resistor, and replace the transistor with another of the same type, again noting the collector voltage.

These three readings show how the bias voltage has been altered by changes in the bias resistor, the load resistor and the transistor respectively. Now set up the bias system shown in Figure 11.6(b) and again adjust the potentiometer until collector voltage is exactly half the supply voltage. Note this value.

As before, connect a 470K resistor in parallel with the bias potentiometer and note the collector voltage. Then remove the 470K resistor, and connect a 6K8 resistor in parallel with the load resistor, noting the effect on the collector voltage. Finally, remove the 6K8 resistor, and find the effect on collector voltage of replacing the transistor. Have the changes in collector voltage been as large as they were in the first case?

Finally, set up the bias system shown in Figure 11.6(c). Measure the collector voltage, then apply the same tests as before and note the effects. Try out several types of silicon transistors in the circuit, noting the collector voltage for each. Does this circuit stabilize bias voltage well?

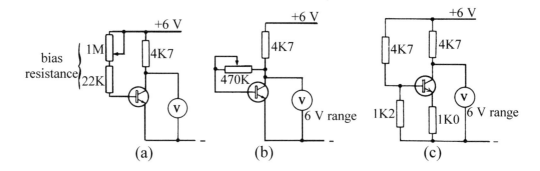

Figure 11.6   Circuits for Activity 11.2

Figure 11.7 illustrates the bias method which is used for a FET in the depletion mode and the circuit commonly used for MOSFET biasing.

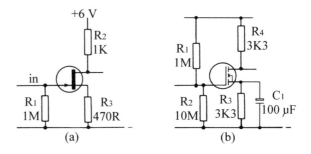

Figure 11.7   Biasing circuit for (a) junction FET, (b) MOSFET

For correct bias, the voltage of the gate for the FET types illustrated here should be negative with respect to the source voltage or, to put the same thing another way, the source voltage must be positive with respect to the gate voltage. In this circuit the positive voltage is obtained from the voltage drop across the resistor R_3 in series with the source. The gate voltage of the JFET is kept at earth (ground level, or zero volts) by the resistor R_1, which needs to have a very large value since no current flows in the gate circuit. The MOSFET circuit uses a potential divider (R_1 and R_2) to maintain a steady voltage on the gate, and the source is biased by the current flowing in R_3. The calculations for the biasing of a FET are considerably simpler than for the biasing of a bipolar transistor.

Question 11.1

Which of the three systems of biasing a transistor is likely to be most affected by replacing a transistor?

Bias failure

Bias failure can be caused by either open-circuit or short-circuit bias components. In any of the circuits shown in Figure 11.5, if a s/c develops across the resistor R_1, the large bias current that will flow in consequence will cause the collector voltage to bottom, and may burn out the base-emitter junction. If the base-emitter junction thus becomes o/c, collector voltage will rise to supply voltage, so that the same fault can be the cause of either symptom. If R_1 becomes o/c, there is no bias supply and the collector voltage cuts off, so that collector voltage equals supply voltage.

In the case of the circuit in Figure 11.5(c), faults in resistors R_2 and R_3 can also affect the bias. Table 11.1 summarizes the possible faults and their effects.

Table 11.1 Bias faults and their effects

Fault	Collector voltage	Emitter voltage
R2 o/c	Low	High
R2 s/c	High	Zero
R3 o/c	High	High
R3 s/c	Low	Zero

Gain and bandwidth

The voltage gain (G) of an amplifier is defined as:

$$G = \frac{\text{Signal voltage at output}}{\text{Signal voltage at input}}$$

with both signal voltages measured in the same way (either both r.m.s. or both peak-to-peak). This quantity G is an important measure of the efficiency of the amplifier and is often expressed in decibels by means of the equation: dB = 20 log G.

Example: Find the gain of an amplifier in which a 30 mV peak-to-peak input signal produces a 2 V peak-to-peak output signal.

Solution: Insert the data in the equation:

$$G = \frac{\text{Signal voltage at output}}{\text{Signal voltage at input}}, \text{ so that } G = \frac{2000}{30} = 66.7.$$

Note that the 2 V must be converted into 2000 mV, so that both input and output signals are quoted in the same units.

Expressing the same answer in decibels:

$$G = 20 \log 66.7 = 36.5 \text{ dB}$$

A decrease in gain from 66.7 to 60 might seem significant, but the same decrease expressed in decibels is only from 36.5 dB to 35.5 dB a change of 1 dB, which is the smallest change of gain that can be detected by the ear when the amplifier is in use. Measurements of gain expressed in decibels can therefore show whether changes of gain are significant or not. Figures of voltage gain by themselves are often misleading for this purpose.

Question 11.2

An untuned amplifier is being tested, and its gain is 3 dB down at 150 kHz. At this frequency, the output signal is 0.25 V for an input of 2 mV. Calculate the gain-bandwidth product.

Frequency response

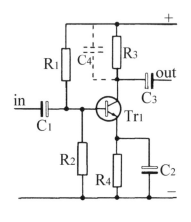

Voltage amplifiers do not have the same value of gain at all signal frequencies. The circuit, left, shows the components in a single-stage transistor amplifier that determine frequency response. (C_4 is shown dotted because it consists of stray capacitances and is not an actual physical component.)

In the circuit, C_1 prevents d.c. from the signal source from affecting the bias at the base of the transistor, and C_3 prevents d.c. from its collector from affecting the next stage. The circuit can therefore give no voltage gain for d.c.; and the amount of gain it can give at low a.c. frequencies is inevitably limited by the action of the capacitors C_1 and C_3. Gain will also be affected by C_2, because this capacitor bypasses the negative feedback action of R_4 for a.c. signals only.

At the high end of the frequency scale, the stray capacitances that are present at the collector of the transistor and in any circuit connected through C_3 are represented by C_4 connected across the load resistor. These strays act to bypass high-frequency signals, so that gain decreases at these frequencies also. Only in the medium range of signal frequencies most commonly used is the gain given by this circuit configuration constant. This constant value is often referred to as *mid-range gain*.

- Note that raising the collector load resistance will provide more gain (if the steady bias current is not reduced) but will decrease the bandwidth.

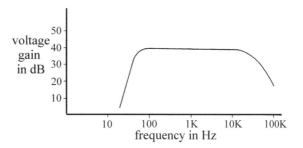

Figure 11.8 Curve of gain plotted against frequency

A typical curve of gain (in decibels) plotted against frequency for such an amplifier is shown in Figure 11.8. In this *frequency-response graph*, note that the frequency scale is logarithmic, so that tenfold frequency steps occupy equal lengths of horizontal scale. This type of scale is necessary to show the full frequency range of an amplifier, and it normally plots gain in dB.

Note that these gain/frequency graphs apply only to sine wave signals. The response of any amplifier to sudden steps in voltage (transients) is affected by the slew rate of the amplifier, a topic that will be dealt with later under the heading of *operational amp*lifiers.

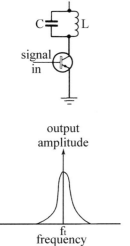

When a tuned circuit is used as the output load of an amplifier, left, the shape of the gain/frequency graph becomes more peaked. The reason is that the tuned circuit presents a high resistance to the signal at the frequency of resonance (f_r). At resonance, this resistance has a value of L/CR ohms, where L is in henries, C in farads, and R is the resistance of the coil expressed in ohms. This value L/CR is called the *dynamic resistance* of the tuned circuit.

- At all other frequencies, the load has a considerably lower value of resistance and acts instead like an impedance with the result that voltage and current become out of phase with one another.

The gain of a tuned amplifier at the resonant frequency is often controlled by an automatic gain control (AGC) voltage. This voltage is applied at the base (or gate of a MOSFET), either opposing or adding to the existing d.c. bias. Reverse ACC uses a d.c. bias voltage that reduces normal bias current; while forward AGC uses a bias voltage that increases the normal bias current. The type of AGC used depends on the type of transistor and the circuit of which it is a component.

To give forward AGC, a resistor is included in series with the load but bypassed by a capacitor, so that signal current does not flow through it and the transistor is so designed that its gain is much lower at low collector voltages. Increasing the bias current then lowers collector voltage and decreases the gain.

Most transistors, however, give greater gain at higher bias currents (given unchanged collector voltage), and so need to use reverse AGC.

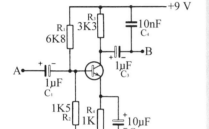

Activity 11.3

Construct the circuit shown here, in which the transistor can be any medium-current NPN silicon transistor such as 2N697, 2N1711, 2N2219 or BFY50. Measure the collector voltage and note its value.

Connect the output of the signal generator to point A, and its earth return to point C. The Y-input of an oscilloscope should also be connected to point A. Switch on all circuits, and adjust the signal generator output to give a 1 kHz, 30 mV peak-to-peak signal at point A.

Now connect the Y-input of the oscilloscope to point B, and read the peak-to-peak value of the output signal. Calculate the gain, convert it to decibels, and record this value as the gain of 1 kHz.

If the output wave is noticeably distorted (i.e., flattened at either peak), reduce the amplitude of the signal generator output until a well-shaped sine-wave appears at the output. Note the new value of input signal needed.

Now multiply your known figure of output voltage at 1 kHz by 0.71, and reduce the output frequency of the signal generator until the output voltage reaches this lower value. Record the signal generator frequency required to achieve this, and call it f_1.

Increase signal generator frequency again until you find a frequency above 1 kHz at which output voltage is reduced by the same amount. Record this frequency as f_2.

Note that during both these readings, the output of the signal generator should remain constant at its original value of amplitude. Check the output amplitude at both frequencies, f_1 and f_2, if there is any doubt.

The factor of 0.71 which you applied above corresponds to a loss of gain of 3 dB. The frequencies f_1, and f_2 are therefore called the lower and the upper 3dB points respectively, and the range of frequencies between them is called the bandwidth of the amplifier.

For an audio amplifier, f_1 and f_2 are quoted, so that such an amplifier can be described as being (for example) *3 dB down at 17 Hz and at 35 kHz*.

In short, the term *bandwidth*, as applied to a tuned amplifier, means the quantity $f_2 - f_1$. A tuned amplifier can therefore be described as having (again for example) *a bandwidth of 10 kHz centred on f_r at 470 kHz*.

The purpose of the 10 nf capacitor C_4 connected across the load resistor R_3, is to ensure that the response of the amplifier at high frequencies will not be too wide for measurement purposes.

Activity 11.4

Still using the circuit for Activity 11.3, observe the changing output waveform while signal generator output at 1 kHz is increased to 300 mV. Sketch the waveform which shows the distortion caused by overloading.

Next, restore the input to its previous value and connect an 8K2 resistor across R_1. Sketch the resulting output waveform, which will show the distortion caused by over-biasing.

With the resistor across R_1, removed, connect a 680R resistor across R_2. Sketch the output waveform, which now shows the distortion caused by under-biasing. Note the changes in voltage readings when the following conditions are present or simulated (bracketed actions):

> base-emitter of o/c (disconnect lead to base)
>
> base-emitter of s/c (short emitter to base connections)
>
> C_2 s/c (short across connections).

The forms of distortion caused by overloading or by faulty biasing will be obvious when the output waveforms are viewed. Note, however, that smaller amounts of distortion caused by curvature of the characteristics are not visible on an output trace, and can only be detected by distortion meters. These filter out the sine-wave which is being amplified, leaving an output which consists only of the distortion, which can then be measured.

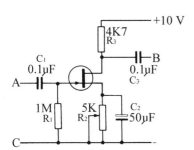

Activity 11.5

Construct the junction FET amplifier shown here. The variable resistor R_2 should be set so that drain voltage is about 7 V when there is no signal output.

Find the gain at 1 kHz, the frequencies which produce the −3 dB points, and note the distortion caused by overloading. Note the changes in voltage readings and (if applicable) gain when the following conditions are present or simulated by removing or short-circuiting components:

- FET faulty (o/c channel or s/c gate to source or drain)
- C_2 o/c
- C_2 s/c
- R_2 high (more than 47K)
- R_3 high (more than 470K).

Table 11.2 below lists faults which have predictable effects, especially on gain and bandwidth.

Table 11.2   Amplifier faults

Fault	Effects
Emitter bypass capacitor o/c	Reduced gain, increased bandwidth
Collector load resistance too low	Low gain, collector d.c. voltage too high
Transistor under-biased	Gain reduced, signal distortion at output

Multiple stage amplification

In most applications, a single transistor is not enough to provide sufficient gain, and several stages of amplification are needed. When an amplifier contains several stages, its total gain, G_t, is given by the equation:

$$G_t = G_1 \times G_2 \times G_3$$

where G_1, G_2, G_3 are the gains of the individual stages.

In decibels, this becomes:

$$\text{Total gain} = (dB)_1 + (dB)_2 + (dB)_3$$

giving the total gain in decibels. Note that the decibel figures of gain are added, whereas the voltage (or current, or power) figures have to be multiplied. This is because logarithms are used in the construction of the decibel figure, and addition of logarithms is the equivalent of the multiplication of ordinary figures.

The coupling together of separate amplifying stages involves transferring the output signal from one stage to the input of the next stage. This can be done in several ways, as described below and illustrated in Figure 11.9 (from which details of all biasing arrangements have been omitted for the sake of clarity). The aim is to transfer the maximum amount of signal from one stage to the next.

Direct coupling involves connecting the output of one transistor to the input of the next, using only resistors or other components which will pass d.c. The result is that both d.c. and a.c. signals will be coupled. A d.c.-coupled amplifier by definition amplifies d.c. signals, so that a small change in the steady base voltage of the first stage will cause a large change in the steady collector voltage of the next. In all d.c.-coupled stages,

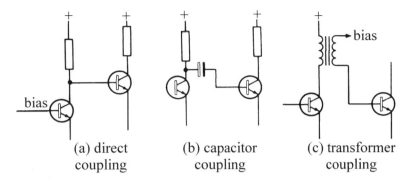

(a) direct
coupling

(b) capacitor
coupling

(c) transformer
coupling

particular attention needs to be paid to bias. A negative feedback biasing system (see later) is usually required.

Figure 11.9   Coupling amplifying stages

Capacitor coupling makes use of a capacitor placed in series between the output terminal of one stage and the input terminal of the next. The effect is that a.c. signals only can be coupled in this way, because d.c. levels cannot be transmitted through a capacitor. When amplifiers need a low −3 dB point when input frequency is only a few Hz, large values of capacitance will be required.

Transformer coupling makes use of current signals flowing in the primary winding of a transformer connected into the collector circuit of a transistor to induce voltage signals in the secondary winding, which in turn is connected to the base of the next transistor. Once again, only a.c. signals can be so coupled; and a well-designed transformer will be needed if signals of only a few Hz are to be coupled. Note that the gain/frequency graph of a transformer-coupled amplifier can show unexpected peaks or dips caused by resonances. For that reason, transformer-coupled amplifiers are seldom used when an even response is of importance.

Question 11.3

An amplifier is supplied with a sine wave input, and an oscilloscope connected to the output shows that the waveform is clipped. What two conditions would you check first to solve the problem?

Negative feedback

Negative feedback can be used in amplifier circuits either to stabilize bias (d.c. feedback) or to stabilize gain (a.c. feedback), or both. The conventional bias circuit illustrated earlier is a form of d.c. negative feedback, and where stages are directly coupled (as in operational amplifiers, see later), d.c. feedback over several stages will be needed to stabilize bias. Faultfinding bias problems can be very difficult in such circuits because every component in the feedback path has an effect on bias. For the moment, however, we shall concentrate on the use of a.c. feedback.

Though it is possible to design single amplifier stages with fairly exact values of voltage gain (it is, for example, quite possible to design an amplifier with a voltage gain of exactly 29 times, if that should happen to be wanted), it is less easy to design multi-stage amplifiers that will give the precise voltage gain required. The reason is that the input and output resistances of transistors vary considerably, depending on the varying h_{fe} values of individual transistors, and that in any form of signal coupling the output resistance of one transistor forms a voltage divider with the input resistance of the next, so attenuating the signal.

The point is simply illustrated in the drawing, left, in which R_1 symbolizes the output resistance of the first transistor and R_2 the input resistance of the second.

A more useful way of designing an amplifier for a specified figure of gain is to aim for one which has too large a value of voltage gain, and then to use negative feedback to reduce this gain to the required figure. One advantage of using negative feedback is that it is often possible to calculate the gain of the complete amplifier without knowing any of the individual transistor gains or resistances. Very often the gain of the amplifier is simply the ratio of two values of fixed resistors. Some feedback methods are shown in Figure 11.10.

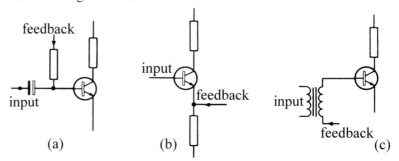

Figure 11.10 Some feedback methods, omitting bias and supply components. (a) Shunt feedback to base, (b) series feedback to emitter, (c) series feedback using a transformer

For negative feedback to be useful, the gain of the amplifier without feedback (called the open-loop gain) must be much greater (100 times or more) than the gain of the amplifier when feedback is applied (the *closed-loop gain*). When this situation exists, the *feedback factor*:

$$\frac{\text{Closed-loop gain}}{\text{Open-loop gain}}$$

becomes important, because distortion and noise generated in the transistors or other components within the amplifier will be reduced by exactly the same ratio. If, for example, open-loop gain is 10 000, and closed-loop gain is 100, then the feedback factor becomes 10 000/100, which is 1/100. This means that the distortion will be reduced to 1/100th of its value in the open-loop amplifier.

Bandwidth, on the other hand, will be increased by the inverse of the factor, which is 100 times in this example, on the assumption that there is nothing else that will limit bandwidth in any other way. You cannot, for example, expect negative feedback to make the bandwidth of an amplifier greater than the bandwidth that the transistor(s) will handle.

There are four principal ways in which negative feedback can be applied to an amplifier, depending on where the signal is taken from and to what point it is fed back.

- A signal fed back from a collector is called *voltage-derived*, because it is a sample of the output voltage signal.

- A signal fed back from a resistor in the emitter circuit is called *current-derived*, because it is proportional to the output current signal and in the same phase as it.

- A signal fed back to a base input is called a *shunt feedback* signal, because the feedback signal is in shunt (or parallel) with the normal input signal.

- A signal fed back to the emitter of an input transistor is called a *series feedback* signal, because the feedback signal is in series with the normal input signal. An alternative method of achieving a series feedback signal is to transformer-couple the feedback to the base of a transistor through the transformer winding, as shown in Figure 11.10(c).

Each of these methods of applying a.c. negative feedback will have the desired effects of reducing gain, noise and distortion, and of increasing bandwidth; but the different methods of connection can affect other features of the complete amplifier. Taking the feedback from the emitter of an output stage, for example, causes the output resistance at the collector of the same transistor to be higher than it would be if the feedback were to be taken from the collector. Alternatively, taking the feedback to a base input causes the input resistance to be lower (often much lower) than it would be if the feedback were to be taken to the emitter.

Some feedback circuits include the input or output resistance of the transistor itself as part of the feedback loop, and are therefore less predictable in action.

Figure 11.11, following, shows two common types of feedback circuit. Figure 11.11(a) uses negative feedback from the emitter of Tr_2 to the base of Tr_1. The feedback is therefore series-derived and shunt-fed. Figure 11.11(b) uses negative feedback from the collector of Tr_2 to the emitter of Tr_1. The feedback is therefore shunt-derived and series-fed. Note that the capacitor C_5 in each case makes the feedback a.c. only, rather than both a.c. and d.c.

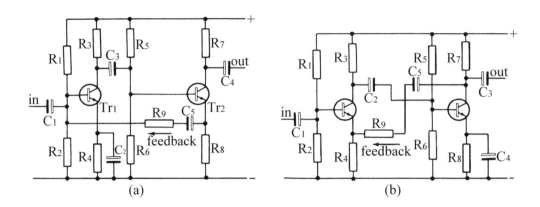

(a)   (b)

Figure 11.11   Two common types of feedback circuit

The table below shows the effect of some component failures on both these circuits.

Table 11.3   Effects of component failures

Fault	Effect
R_9 or C_5 o/c	No negative feedback, high gain, possible instability
S/c across R_9	Greatly reduced gain
C_5 s/c	Bias of Tr_1 incorrect

Activity 11.6

Construct the circuit shown here. Both transistors can be any general-purpose silicon NPN transistor. Then observe the effects of two different types of feedback.

(a) Series-derived, shunt-fed. Measure the voltage gain of the complete amplifier. (Note that a small input signal should be used to prevent overloading). Now remove C_4, so introducing negative feedback through R_1. Measure and note the new value of voltage gain.

(b) Shunt-derived, series-fed. Replace C_4, remove C_2, and again measure the value of circuit gain. This gain is now the loop gain of the feedback loop through R_2. Note its value and compare with (a) above. What form of feedback is applied through R_2?

Question 11.4

An amplifier suddenly displays a large increase in gain and distortion. Where would you look for a cause of this problem?

- Note that negative feedback cannot improve the performance of an amplifier for pulse waveforms if the slew rate of the amplifier is being exceeded. This topic will be dealt with later as part of the limitations of operational amplifiers.

Differential amplifiers

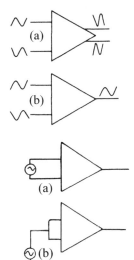

The balanced or differential, amplifier circuit is much used in industrial equipment, and is the basic circuit of most linear ICs. A balanced amplifier has two inputs, one the phase-inverted 'image' of the other. The output often (but not always) appears in similar balanced form. The drawing, left, shows an amplifier with balanced output (a), and with unbalanced output (b). Balanced output signals are said to be *balanced about earth* if the sum of the output voltages at any instant is always zero.

The value of balanced signal amplifiers lies in the fact that such amplifiers make it possible to amplify small (balanced) signals even in the presence of very large (unbalanced) noise pulses, so that the amplifier output contains only the wanted signal. Thus ideally there should be no output signal for any unbalanced input, called a *common mode input signal*. The comparison of these two inputs is shown in the lower set of drawings with balanced input (a) and common-mode input (b).

The ratio of amplifier voltage gain for each type of signal is called the *common mode rejection ratio*, and is generally expressed in decibels by means of the following formula:

$$\text{Common mode rejection ratio} = 20 \times \log \frac{G_b}{G_{cm}}$$

where G_b = voltage gain of amplifier for balanced signals, and G_{cm} = voltage gain of amplifier for unbalanced signals. Alternatively stated, the CMRR is the ratio of the input signal levels needed to produce the same level of output signal in each case. In this context, the phrase *common-mode* means an unbalanced signal that appears at both input terminals in the same phase.

Values of common-mode rejection ratio of 100 dB and more can be obtained by suitable amplifier design, which means that voltage gain for the balanced signal is 100 000 times, or more, the voltage gain for the unbalanced signal. Such large values of common-mode rejection ratio render industrial amplifiers virtually insensitive to interference, provided only that the wanted signal is in balanced form.

This wanted signal is generally derived from some sort of transducer, and many types of transducers can be connected so that they give balanced

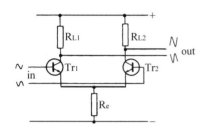

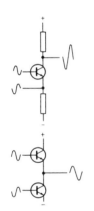

output signals. The various types of bridge connection, for example, can all be arranged to produce a balanced signal; and inductive transducers can be wound with a centre-tap so that their outputs are always balanced.

The basic balanced amplifier circuit, known either as the *differential amplifier* or *long-tailed pair*, is shown in the drawing with the biasing circuits omitted for the sake of clarity. The resistor R_e is an essential part of the circuit because, when the circuit is correctly biased and operated, the signal voltage across it is always zero. An increase in (instantaneous) voltage at the base of one of the transistors ought to be accompanied by an identical decrease at the base of the other. A signal will therefore appear across R_e only when the inputs are not perfectly in anti-phase.

R_e therefore forms part of the action which discriminates against common-mode signals. It is the *tail* of the long-tailed pair and intended to pass a constant current which is shared between Tr_1 and Tr_2.

In a balanced amplifier, the signal voltage is applied equally between the two base inputs, and the amplified output signal is taken from between the collectors. Given that the transistors have identical values of mutual conductance (they form a matched pair), any signal which appears identically at both bases will give rise to an amplified and inverted signal at both collectors; but the voltage between the collectors will be unaltered.

It is this action which provides the rejection of the noise signals a large enough value of R_e ensuring that any common-mode signal arising from a lack of balance between the transistors is offset by negative feedback from the signal voltage across R_e. Hum or other interference on the supply line is a common-mode signal, and so produces no output.

The output of a differential amplifier can be used to drive another similar stage, or it can be converted to unbalanced form by using one of the circuits shown, left, with bias circuitry again omitted.

The differential amplifier is also a versatile circuit for the handling of unbalanced input signals. If one transistor base in the circuit is connected (by a capacitor, for example) to signal earth, an unbalanced signal applied to the base of the other transistor will result in a balanced signal between the two collectors (Figure 11.12).

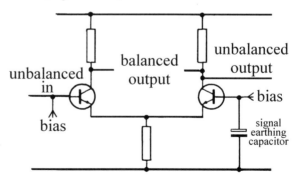

Figure 11.12   Unbalanced input, output either balanced or unbalanced

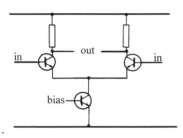

The voltage gain is approximately equal to that of a single transistor operating at the same bias current. A single-ended output can also be obtained, but the voltage gain will then be restricted to half what it would have been if the output had been a balanced one from the same circuit. Input resistance is high in both these circuits, and the emitter resistor acts as the coupling between the two transistors. The circuit of Figure 11.12 is particularly useful in practice, for even with an unbalanced input the circuit discriminates against interference pulses and supply hum affecting both bases alike.

Note that in Figure 11.12 the output signal is in phase with the input signal. A significant improvement in performance can be obtained if the resistor R_e is replaced by another transistor, so biased as to keep the current flowing into the differential pair constant. This circuit variation (shown here with the biasing components again omitted) is much used.

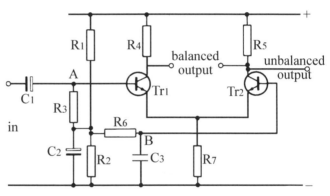

Figure 11.13   Circuit for Activity 11.7

Activity 11.7

Assemble the long-tailed pair circuit shown in Figure 11.13 using the following component values: $R_1 = 12K$; $R_2 = 2K2$; R_3 and $R_6 = 56K$; R_4 and $R_5 = 6K8$; $R_7 = 1K$; C_1, C_2 and $C_3 = 10$ µF, 25 v.w. Tr_1 and Tr_2 can be any small-signal silicon NPN voltage amplifier transistors.

Connect the circuit shown to a power supply in which neither positive nor negative line is earthed. Measure the collector and emitter voltages relative to the negative line, and note the values. Connect the inputs to an unearthed signal generator output and the differential outputs to an oscilloscope. Note that if either the power supply or the signal generator has one lead earthed, the signals will be shorted out.

• Electrical safety requirements are most easily met if both the circuit and the signal generator are battery-operated. If an oscilloscope with a differential input is available, an earthed signal generator can be used provided that a battery supply is used for the differential amplifier.

Measure the gain at 400 Hz, and note this value as the value of differential gain, G_b. Now break the circuit at point B, and reconnect the base of Tr_2 to point A. Again measure the gain at 400 Hz. The result will give the common-mode gain, G_{cm}. Calculate the common-mode rejection ratio in decibels.

Re-connect the oscilloscope and measure the output signal between the collector of Tr_2 and the negative line. A single-ended output is now being taken. Re-connect Tr_2 base to point B, find and note the gain at 400 Hz. Now break the circuit again at point B and connect the base of Tr_2 to point A once more. Find the value of gain for this arrangement. Is there any significant common-mode rejection when a single-ended output is used?

Operational amplifiers

Operational amplifiers (commonly known as opamps) were originally developed to perform the equivalent of mathematical operations in analogue computers. Until integrated circuits were invented, however, their use in discrete form was very limited. Once mass production of integrated circuits started, opamps were found to have many very useful properties for other applications. As a result opamps are now found in a very wide variety of equipment.

IC operational amplifiers are multi-stage differential amplifier units with direct coupling and very large values of gain. The open-loop gain is the amount of gain that would be obtained by using the IC as an amplifier without feedback, but this amount is so large (usually 100 dB or more, corresponding to voltage gain of 100 000 or more) as to be unusable. IC amplifiers are always used with feedback, and since direct coupling is normally used, the feedback is used both to establish bias and to establish signal gain.

- Increased reliability due to a construction that leads to fewer interconnections to develop dry joints.
- Reduced size compared with their discrete component counterparts.
- Reduced costs with volume production.
- Faster operation or better high-frequency response due to shorter signal path lengths.
- IC fabrication gives a better control over the spread or variation of device parameters.
- Reduced assembly time due to fewer soldered joints.
- Since the final performance of the circuit can be closely controlled by using negative feedback, the overall design of circuits is simpler.

A typical opamp example, now quite old, is the 741. The remainder of this section will describe the operation of the IC version of this circuit.

The ideal opamp would have the following characteristics:

- very high input resistance,
- very low output resistance,
- very high voltage gain, open-loop, and
- very wide bandwidth.

The following table shows typical characteristics for a 741 opamp. You can see that the first three of the above requirements are easily met; while the bandwidth, though small by audio standards, is wide enough for the signals that the 741 was designed to handle, signals ranging from d.c. to a few hundred Hz.

Table 11.4 741 opamp characteristics

Input resistance	Output resistance	Open-loop gain	Bandwidth	Max. supply voltage	Max. load current
1M	150R	100 dB	1 kHz @ 60 dB gain	±18 V	10 mA

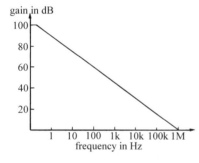

gain in dB

frequency in Hz

The gain-frequency response of a typical opamp is shown here, left. Remember that by using negative feedback the stage gain of an amplifier is reduced but at the same time its bandwidth is increased. This is due to the fact that the amplifier's gain bandwidth product is a constant. This feature of being able to trade gain for bandwidth is probably the most important feature of the opamp and forms one of the design steps in the production of a practical amplifier, but care is needed because there are limits imposed by slew rate, as we shall see.

The frequency range of an opamp depends on two factors, the gain-bandwidth product for small signals, and the slew rate for large signals. The gain-bandwidth product is the quantity, $G \times B$, with G equal to voltage gain (not in dB) and B the bandwidth upper limit in Hz. For the 741, the GB factor is typically 1 MHz, so that, in theory, a bandwidth of 1 MHz can be obtained when the voltage gain is unity, a bandwidth of 100 kHz can be attained at a gain of ten, a bandwidth of 10 kHz at a gain of 100 times, and so on. This trade-off is usable only for small signals, and cannot necessarily be applied to all types of operational amplifiers. Large amplitude signals are further limited by the slew rate of the circuits within the amplifier. The slew rate of an amplifier is the value of change of output voltage:

$$\text{slew rate} = \frac{\text{maximum voltage change}}{\text{time taken}}$$

and the units are usually volts per microsecond. Because this rate cannot be exceeded, and feedback has no effect on slew rate, the bandwidth of the opamp for large signals, sometimes called the power bandwidth, is less than that for small signals. The slew rate limitation cannot be corrected by the use of negative feedback; in fact negative feedback acts to increase distortion when the slew rate limiting action starts, because the effect of the feedback is to increase the rate of change of voltage at the input of the amplifier whenever the rate is limited at the output. This accelerates the overloading of the amplifier, and can change what might be a temporary distortion into a longer-lasting overload condition.

Slew rate limiting arises because of internal stray capacitances which must be charged and discharged by the current flowing in the transistors inside the IC, so that improvement is obtainable only by redesigning the internal circuitry. The 741 has a slew rate of about 0.5 V/μs, corresponding to a power bandwidth of about 6.6 kHz for 12 V peak sine wave signals. The slew rate limitation makes opamps unsuitable for applications which require fast-rising pulses, so that a 741 should not be used as a signal source or feed (interface) with digital circuitry, particularly TTL circuitry, unless a Schmitt trigger stage is also used. Higher slew rates are obtainable with more modern designs of opamps; for example, the Fairchild LS201 achieves a slew rate of 10 V/μs.

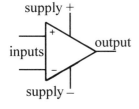

The circuit symbol for an opamp is illustrated here. Two power supplies are needed (though single supply lines can also be used by modifying the circuits). The dual supply is balanced about earth. The opamp has two inputs, labelled (+) and (−) respectively, and a single output. The internal circuit is that of a balanced amplifier, so that the output voltage is an amplified copy of the voltage difference between the two inputs. The (+) sign at one input indicates that feedback from the output to this input will be in phase, positive. The (−) sign at the other input indicates that feedback to this input will be out-of-phase, negative.

The voltage gain, open-loop, is some 10^5 (100 dB) or more. Since the internal circuit is completely d.c. coupled, both d.c. and a.c. negative feedback will be needed for linear amplifier applications.

An operational amplifier, unless the bias is stabilized, will suffer from *drift* and *offset*. Consider an opamp with equal positive and negative power supplies. If both inputs are at the same zero-voltage level, you would expect the output also to be at zero (ground) level. In fact, it will not be, and the difference between input levels needed to obtain a zero output is called the offset. It might be possible to apply inputs that ensured a steady output (even if not zero), but in such conditions the output voltage would vary with temperature and time, the effect called drift. The problems of offset and drift are dealt with by using d.c. feedback bias circuits. See later for more details of drift and offset in practical circuits.

Opamp circuits

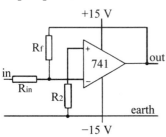

The diagram, left, illustrates a basic phase-inverting amplifier circuit. Balanced power supplies are used, with the (+) input earthed and the (−) input connected to the output by a resistor R_f which provides negative feedback of both d.c. and signal voltages. R_{in} serves to increase the input resistance. This inverting voltage amplifier circuit is the most common application for an opamp.

Although the (−) input is not actually connected to earth, its voltage (either d.c. or signal) is earth voltage, and it is said for this reason to be a *virtual earth*. What happens is this. The (+) input is earthed, and any voltage difference between the inputs is amplified. Suppose, for example, that the (−) input is at 0.1 mV below earth voltage. This 0.1 mV will be amplified by 10^5, giving a +10 V output (note the phase change). This voltage then drives current through R_f to increase the voltage at the (−) input until it equals the voltage at the (+) input.

Similarly, a 0.1 mV positive voltage at the (−) input would cause the output to swing to −10 V, again causing the input to return to zero. Two points should be noted:

> The voltage gain is so high that the assumption that the (−) input is at zero voltage is justified.
>
> Any unbalance in the internal circuit will cause an offset voltage at the input, so that the (−) input may need to be at a very small positive or negative voltage to maintain the output at zero (the input offset voltage).

This input offset can be reduced in the Type 741 by adding the offset balancing resistor shown in the circuit here, in which the small figures are the IC pin numbers. The size of the offset voltage will, however, vary as the temperature of the IC varies, so that some compensation may be needed in circuits such as high-gain d.c. amplifiers, in which offset can be troublesome. This is known as the input offset voltage temperature drift usually abbreviated to drift. The offset is not a problem when large amounts of d.c. negative feedback are used.

Due to the virtual earth at the (−) input of the basic phase-inverting circuit the circuit input resistance becomes equal to R_{in}. It can also be shown that the stage gain is simply the ratio R_f/R_{in}. To minimize the effects of temperature drift, the resistor R_2 is made equal to the parallel combination of R_f and R_{in}, that is,

$$\frac{R_f \times R_{in}}{R_f + R_{in}}$$

Again due to the virtual earth, several input signals can be applied to the (−) input through resistors. This forms the basis of the summing amplifier, where the input signals are added.

The drawing, left, shows a type of non-inverting circuit of the general type known as a *follower*. In this circuit the signal input is applied to the (+) input, whose d.c. value is normally fixed at earth voltage by R_1. The feedback resistor R_2 ensures that the (−) input is a virtual earth. A signal coming into the input will now produce an identical signal, in phase, at the output. The input resistance of the circuit is extremely high, the output resistance very low, and the circuit is used mainly to obtain these characteristics (like the emitter-follower transistor amplifier).

Unlike the familiar emitter-follower, this type of circuit can be modified to produce voltage gain, as is shown here, left. Such a circuit should only be used when the signal input is of small amplitude because the effect of the feedback is to make the input signal a common-mode signal, and the amplitude of common-mode signals needs to be kept below a specified value in most opamp designs. The gain in the example is given by:

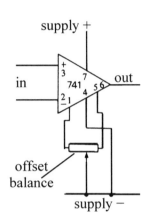

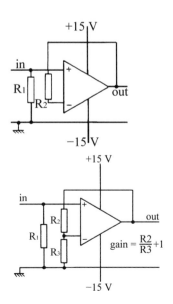

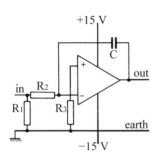

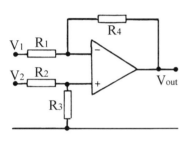

$$G = \frac{R_2}{R_3} + 1$$

The characteristics of the opamp make it possible to build integrating circuits giving much better performance than can be obtained using only passive components.

The illustration, left, shows a simple opamp integrator which functions also as a low-pass filter. The time constant CR_2 seconds should be approximately five times the periodic time t of the input signal. A fully practical circuit, however, must also include some method of setting the output voltage to zero before the circuit is used, particularly if the d.c. level of the output is important. If the integrator is to be used for a.c. only, a bias resistor (such as R_1) will serve to prevent drift.

If C and R_2 of the integrator example are interchanged and their time constant is approximately one fifth of the periodic time t, the circuit becomes a differentiator. The output waveform then depends on the rate of change of the input signal. For input signals of a different shape, the differentiator acts as a high-pass filter.

When an opamp is operated in the differential mode as drawn (with bias omitted) here, its output V_{out}, is proportional to the difference between the two inputs V_1 and V_2. The polarity of V_{out} depends upon which input is the larger. If V_1 is the greater, then V_{out}, is negative.

Figure 11.14 shows positive feedback used in an opamp circuit. This has the effect of providing a Schmitt trigger action with hysteresis. If we imagine the input voltage starting low, then with input voltage rising, the output will suddenly switch over (high). As the input voltage falls, the output voltage will switch back again, but not at the same voltage as caused the switch in the other direction.

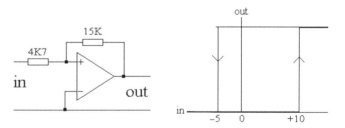

Figure 11.14   Opamp providing Schmitt trigger action

Opamps can thus be used in many roles, the following being typical:

- As differential amplifiers, having exceptionally good common-mode rejection ratios, and with response down to d.c.

- As amplifiers for audio frequencies whose gain is controlled entirely by the values of the feedback components.

- As high-pass and low-pass active filters.

- As integrators and differentiators for the shaping of signal waveforms.

- As sensitive switching circuits, triggered by the very small size of the differential input voltage needed to switch the output from one level to another.

- As comparators, providing an output which is an amplified version of the voltage difference between two signal inputs (a.c. or d.c.). This application can be extended to provide a Schmitt trigger action with hysteresis.

Activity 11.8

Construct the two opamp circuits shown in Figure 11.15 (which are both shown, for convenience, as operating from a single-ended power supply). In each circuit, measure the d.c. voltages with a high-resistance meter, both at the output of the 741 and at each input. Then, with the aid of a signal generator and an oscilloscope, measure the amount of gain at 400 Hz.

Does the amount of gain shown correspond to what you would expect to get from the values of the resistors in the two circuits?

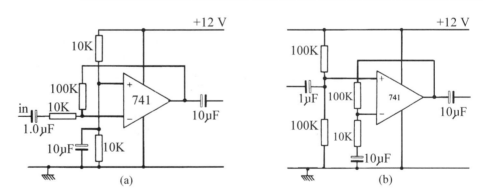

Figure 11.15 Circuits for Activity 11.8

Activity 11.9

Connect up the circuit in Figure 11.16 which uses the 741 opamp as a differential amplifier. The input arrangement makes it possible to apply signals of differing amplitudes to the two inputs, and the use of feedback resistor R_4 maintains the gain at a low value.

(a) Using the values as shown in the diagram, set the signal-generator and the potentiometers so that an output of about 2 V p–p is obtained. Measure the peak-to-peak signal voltage levels at the two inputs to the 741, pins 2 and 3. Write your measurements in the form of a table and calculate the differential gain.

(b) Replace R_2 and R_4 by 820K resistors and repeat your measurements. What value of gain does this change cause?

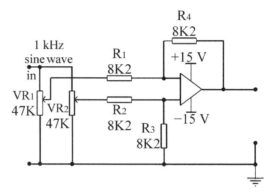

Figure 11.16 Circuit for Activity 11.9

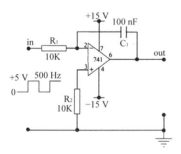

Activity 11.10

Connect up the opamp integrator circuit as shown. Arrange for a square-wave input of 5 V peak-to-peak at 500 Hz from the signal generator. (a) Using the circuit as shown, measure the peak-to-peak output voltage, and sketch the output waveform. (b) Change the value of C_1 to 220 nF and repeat your measurements.

Question 11.5

Square-wave testing is sometimes used as quick way of assessing an audio amplifier. Give one good reason why this might be misleading.

Power amplifiers

Both voltage amplifiers and current amplifiers play important parts in electronic circuits, but not all of them are capable of supplying some types of load. For instance, a voltage amplifier with a gain of 100 times may be well able to feed a 10 V signal into a 10K load resistance, but quite incapable of feeding even a 1 V signal into a 10 ohm load. A current amplifier with a gain of 1000 may be excellent for amplifying a 1 μA signal into a 1 mA signal, but unable to convert a 1 mA signal at 10 V into a 1 A signal at the same voltage.

The missing factor common to both these examples is power. A signal of 1 A (r.m.s.) at 10 V (r.m.s.) represents a power output of 10 W, and the small transistors which are used for voltage or current amplification cannot handle such levels of power without overheating.

Transistors intended to pass large currents at voltage levels of more than a volt or so must have the following characteristics:

- they must have low output resistance,
- they must have good ability to dissipate heat.

A low output resistance is necessary because a transistor with a high output resistance will dissipate too much power when large currents flow through it. Low resistance is achieved by making the area of the junctions much larger than is normal for a small-signal transistor.

The ability to dissipate heat is necessary so that the electrical energy that is converted into heat in the transistor can be easily removed. If it were not, the temperature of the transistor (at its collector-base junction) would keep rising until the junctions were permanently destroyed.

Given suitable transistors with large junction areas and good heat conductivity to the metal case, the problem of power amplification becomes one of using suitable circuits and of dissipating the heat from the transistor.

In practice, most power amplifier stages are required to provide mainly current gain, since the voltage gain can be obtained from low-current stages before the power-amplifier stage. The power amplifier stage can therefore have very low voltage gain or even attenuate the voltage level.

Heat sinks (reminder)

Heat sinks take the form of finned metal clips, blocks or sheets which act as convectors passing heat from the body of a transistor into the air. Good contact between the body of the transistor and the heat sink is essential, and silicone grease (also called heat-sink grease) greatly assists in this contact.

Many types of power transistor have their metal cases connected to the collector terminal. It is therefore necessary to insulate them from their heat sinks. This is done by using thin mica washers between the transistor and its heat sink, with insulating bushes inserted on the fixing-bolts in addition. Mica is an electrical insulator and a heat insulator, but for a thin washer the heat conductivity can be reasonably good. Heat-sink grease should always be smeared on both sides of all such washers.

Classes of bias

Several different methods exist for biasing transistors which are to be used in power output stages. Class A and Class B are the names used for two types of commonly used biasing systems for audio, and Class C is used for RF transmitter amplifiers.

The transistor in a Class A stage is biased so that the collector voltage is never bottomed, nor is current flow ever cut off. Output current flows for the whole of the input cycle. It is the bias system that is used for linear voltage amplifiers. Class A operation of a transistor ensures good linearity, but suffers from two disadvantages:

- A large current flows through the transistor at all times, so that the transistor needs to dissipate a considerable amount of power.

- This loss of power in the transistor inevitably means that less power is available for dissipation in the load, and a Class A stage can seldom be more than about 30 per cent efficient, meaning that the a.c. power can seldom be more than 30 per cent of the d.c. power.

Even in an ideal Class A amplifier, with the load and amplifier output resistances matched, only some 50 per cent of the available power would be delivered to the load, with the remaining 50 per cent being dissipated by the amplifier in the form of heat. When the resistances are mismatched, the power transfer ratio is even lower.

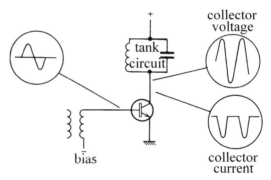

Figure 11.17   A Class B radio-frequency amplifier

In a Class B amplifier (see Figure 11.17), the power transistor conducts for only one-half of the duration of the input sine wave. A single transistor is therefore unusable unless the other half of the wave can be obtained in some other way. At radio frequencies this can be done by making use of a load which is a resonant circuit (called a *tank-circuit*). The tank circuit is made to oscillate by the conduction of the transistor, and the action of the resonant circuit continues the oscillation during the period when the transistor is cut off.

This principle can only be applied, however, if the signal to be amplified is of fairly high frequency. This is because the values of inductance and capacitance needed to produce resonance at, say, audio frequencies would be too large to be practical; and because a load of such a size would in any case too greatly restrict the bandwidth.

An alternative method which is used at audio and other low frequencies is to use two transistors, each conducting on different halves of the input wave. Such an arrangement, illustrated left, is called a *push-pull* circuit.

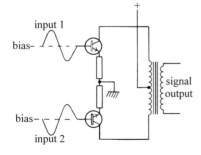

Push-pull circuits can be used in Class A amplification, and are essential for use in Class B.

Class B stages possess the following advantages over Class A:

- Very little steady bias current flows in them, so that the amplifier has only a negligible amount of power to dissipate when no signal is applied.
- Their theoretical maximum efficiency is 78.4 per cent, and practical amplifiers can achieve efficiencies of between 50 per cent and 60 per cent. This means that more power is dissipated in the load itself, and less wasted as heat by the transistors.

The disadvantages of Class B compared to Class A are that:

- The supply current changes as the signal amplitude changes, so that a regulated supply is often needed.
- More signal distortion is caused, especially at that part of the signal where one transistor cuts off and the other starts to conduct. This part of the signal is called the cross-over region.

Typical circuits

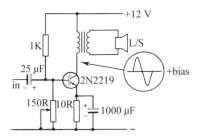

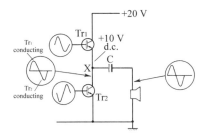

The circuit here illustrates a single-transistor power amplifier biased in Class A. Because no d.c. can be allowed to flow in the load (which is in this case a loudspeaker), a transformer is used to couple the signal from the transistor to the load.

With no signal input, the steady current through the transistor is about 50 mA and the supply voltage is 12 V. Because of the low resistance of the primary winding of the transformer, the voltage at the collector of the transistor is about equal to the supply voltage.

When a signal is applied, the collector voltage swings below supply voltage on one peak of the signal and the same distance above the supply voltage on the other peak of the signal. Because of this transformer action the average voltage at the collector of the transistor remains at supply voltage when a signal is present.

In a Class A stage of this type the current taken from the supply is constant whether a signal is present or not. Any variation in current flow from no-signal to full-signal conditions indicates that some non-linearity, and therefore distortion, must be present.

A circuit that is often used for Class B amplifiers is illustrated, left. It is known as the single-ended push-pull or totem-pole circuit. Two power transistors are connected in series, with their mid-connection (point X in the figure) coupled through a capacitor to the load which may be either a loudspeaker, the field coils of a TV receiver, or the armature of a servomotor. During the positive half of the signal cycle Tr_1 conducts, so that the output signal drives current through the load from X to ground. During the negative half of the cycle, Tr_1 is cut off and Tr_2 conducts, so that the current now flows in the opposite direction, through the load and Tr_2.

The coupling capacitor C forms an essential part of the circuit. When there is no signal input, C is charged to about half supply voltage, i.e. the voltage at point X. The voltage swing at this point X, from full supply voltage in one direction to ground or zero voltage in the other, thus becomes at the load a voltage swing of the same amplitude centred round zero volts.

Bias and feedback

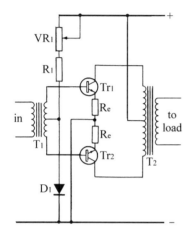

Class A output stages need a bias system that will keep the standing quiescent current (i.e. the d.c. current with no signal input) flowing through them almost constant, despite the large temperature changes caused by the standing current. One common method is to use a silicon diode as part of the bias network. The diode should be a junction diode attached to the same heat sink as the power transistor(s).

As a silicon junction is heated, the junction voltage (about 0.6 V at low temperatures) which is needed for correct bias becomes less. A fixed-voltage bias supply composed of resistors would therefore over-bias the transistor as the temperature increased. A silicon diode compensates for the change in the base-emitter voltage of the transistor, since the forward voltage of the diode is also reduced as its temperature rises.

The Class A push-pull stage in the illustration uses two silicon transistors coupled to the load by a transformer. The input signal also is coupled to the stage by a transformer, which ensures that both transistors obtain the correct phase of signal. The use of a second transformer in such a circuit is often undesirable because it reduces the bandwidth and makes the amplifier less stable if negative feedback is used (because of the unpredictable changes of phase that occur in a transformer at extremes of frequency). A phase-splitter stage that uses transistors rather than a transformer can be used. Such a phase-splitter can use a transistor with equal loads in collector and emitter, or a long-tailed pair circuit with unbalanced input and balanced output.

The bias current for this circuit is taken to the centre-tap of the secondary winding of the phase-splitter (or driver) transformer T_1, and some additional stability is obtained by using the negative feedback resistors R_e in the emitter leads. Any increase in the bias current of the power transistors will cause the emitter voltage of both transistors to rise, so reducing the voltage between base and emitter and thereby giving back-bias to the input.

In both the circuits that have been described so far the bias is adjusted so that the correct amount of steady bias current flows in the output stage. To adjust this bias current in the single-transistor stage requires breaking the collector circuit to connect in a current meter, and then adjusting VR_1 so as to give the correct current reading. The push-pull circuit can be set by measuring the voltage across the emitter resistors, R_e, and by adjusting VR_1 to give the correct value of voltage at these points.

A more elaborate circuit using the Class AB complementary stage is shown in Figure 11.18. In Class AB, each amplifier is biased to a value lying between those appropriate for either Class A or Class B. The output current in each stage thus flows for slightly more than half of each input cycle. The effect is to minimize the cross-over distortion that occurs in Class B operation when both transistors are non-conducting.

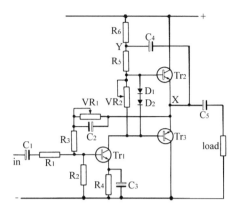

Figure 11.18 A complementary output stage of the totem-pole type with driving stage and feedback connections

In this circuit, both a.c. and d.c. feedback loops are used. The d.c. feedback is used to keep the steady bias current at its correct low value, and the a.c. feedback is used to correct the distortions caused by Class B operation, in particular the cross-over distortion.

Two bias-adjusting settings are needed in this circuit. The potentiometer VR_1 sets the value of the bias current in Tr_1, so that the amount of current flowing through resistors R_2, R_3 and VR_2 is controlled. The potentiometer is adjusted so that the voltage at point X is exactly half the supply voltage when there is no signal input. Potentiometer VR_2 controls the amount of current passing through the output transistors.

This type of output circuit is called a complementary stage, because it uses complementary transistors, one NPN and one PNP type, both connected as emitter followers. With the emitters of both Tr_2 and Tr_3 connected to the voltage at point X, both output transistors are almost cut off when no signal is present. When VR_2 is correctly adjusted, the voltage drop between points A and B is just enough to give the output transistors a small standing bias current (2 to 20 mA) to ensure that they never cut off together. The value of this steady current is usually set low so as to keep cross-over distortion to a minimum – the actual value will be that recommended by the manufacturers. A current meter must be used to check the value of bias current. Another name for this circuit is the *Lin circuit*, named after its inventor.

You can see that there are two a.c. feedback loops in this amplifier. Negative feedback, to improve linearity, is taken through C_2, and positive feedback is taken through C_4 to point Y. This positive feedback, sometimes called *boot-strapping*, cannot cause oscillation because the feedback signal is always of smaller amplitude than is the normal input signal at that point (an emitter follower has a voltage gain slightly lower than unity). It has, however, the desirable effect of decreasing the amount of signal amplitude that is needed at the input to Tr_2.

Impedance matching

The ideal method of delivering power to a load would be to use transistors which had a very low resistance, so that most of the power (I^2R) was dissipated in the load. Most audio amplifiers today make use of such transistors to drive 8 ohm loudspeaker loads.

For some purposes, however, transistors that have higher resistance must be used or loads that have very low resistance must be driven, and a transformer must be used to match the differing impedances. In public address systems, for example, where loudspeakers are placed at considerable distances from the amplifier, it is normal to use high-voltage signals (100 V) at low currents so as to avoid I^2R losses in the lines. In such cases the 8-ohm loudspeakers must be coupled to the lines through transformers.

It was shown in Chapter 2 that the turns ratio:

$$N = \sqrt{\frac{\text{output impedance of amplifier}}{\text{impedance of load}}}$$

provides the maximum transfer of power.

Activity 11.11

Connect the power amplifier illustrated here (or any other suitable amplifier) to an 8-ohm load resistor, and adjust the signal generator to give a 2 V peak-to-peak output across it. Vary the frequency of the signal generator, first down to the frequency at which the voltage output is 1.4 V (0.707 × 2 V), or 3 dB down, and then upwards to the frequency at which the output is again 1.4 V, or 3 dB up. Note these 3 dB frequencies.

You can see that at each of these 3 dB frequencies the power output of the amplifier is half as much as it is at 400 Hz (0.707 × 0 707 ≅ 0.5). The frequencies are therefore known as the half-power points. Note the power bandwidth of the amplifier, which is the frequency range between the half-power points.

Question 11.6

What two adjustments are needed to set the operating conditions of a totem-pole type of Class AB amplifier stage?

Faults

Faults in output stages are usually caused by over-dissipation of power, which in turn can be caused by over-loading or over-heating. An output stage can be over-loaded, when capacitor coupling is used, by connecting a load having too low a resistance. The usual result is to burn out the output transistor(s).

In most Class AB totem-pole circuits, even the most momentary short circuit at the output (caused for example, by faulty connections) will cause the transistor that is connected to the + supply line (Tr_2 in Figure 11.18) to burn out if a signal is being amplified.

Excessive bias currents can often be traced to the failure of a diode in the bias chain, or to a burn-out of the biasing potentiometer in a totem-pole circuit.

Unexpected clipping of the output can be caused by failure of the bootstrap capacitor (C_4 in Figure 11.18), or by a fault in the bias resistors which has caused the voltage at point X to drift up or down.

IC power amplifiers are now widely used, replacing power stages which use separate (discrete) transistors. Unlike opamps, however, power amplifier ICs take no standardized form, though many use the same scheme of connections. Figure 11.19 shows a circuit diagram (courtesy of RS Components Ltd.) for the TDA2030 power amplifier used with a single 15 V power supply. The IC is manufactured on a steel tab which can be bolted to a heatsink for efficient cooling.

The circuit is very similar to that of a non-inverting opamp voltage amplifier, but the practical layout of the circuit has to be designed so that the decoupling capacitors (C_2 and C_3) are very close to the IC. The electrolytic capacitor C_3 needs to be bypassed by a plastic dielectric capacitor because the impedance of an electrolytic capacitor rises at the higher frequencies. At these frequencies, C_2 performs the decoupling in place of C_3. The series circuit of R_6 and C_6 is designed to maintain stability and is called a Zobel network – it is also widely used in discrete transistor amplifiers. This circuit will deliver up to 13 W of audio output into the 4 ohm loudspeaker.

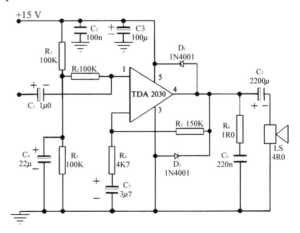

Figure 11.19 A power amplifier circuit, courtesy of RS Components Ltd., using the TDA 2030 IC

Note that the circuits used in hi-fi amplifiers are usually of a specialized nature, and many use power MOSFETs in the output stages. Some types can be serviced, using information from the manufacturers, but others should be returned to the manufacturer for servicing, particularly if there has been a failure of any critical part. For example, failure of one power MOSFET will require replacement with a matching unit, and few servicing workshops will have equipment for precisely measuring power MOSFET parameters.

Answers to questions

11.1   The single resistor (base to supply) type.

11.2   18.75 MHz.

11.3   Check for excessive input signal, then for incorrect bias.

11.4   The NFB loop is broken, or resistance values in the loop have changed.

11.5   The response to a square wave might be affected by the slew rate of opamps in the circuit.

11.6   The d.c. level of the output and the quiescent current through the output transistors.

12 Oscillators and waveform generators

Positive feedback and oscillation

The effects of negative feedback of a.c. signals on a voltage amplifier were described in Chapter 11. Remember that negative feedback is achieved by subtracting a fraction of the output signal from the input signal of an amplifier. In practice, this is done by adding back the feedback signal in antiphase, so that feedback from an output which is in antiphase to an input is always negative unless some change of phase occurs in the circuit used to connect the output to the input (see later under RC oscillators).

If a signal which is *in phase* with the input is fed back, the feedback becomes positive. Positive feedback takes place when a fraction of the output signal is added to the signal at the input of an amplifier so increasing the amplitude of the input signal. The result of positive feedback is higher gain (though at the cost of more noise and distortion) if the amount of feedback is small. If the amount of feedback is large, the result is oscillation.

An amplifier oscillates when:

- the feedback is positive at some frequency; and
- the voltage gain of the amplifier is greater than the attenuation of the feedback loop (see Chapter 11 again to remind yourself about loop gain).

If, for example, 1/50th of the output signal of an amplifier is fed back in phase, oscillation will take place if the gain (without feedback) of the amplifier is more than 50 times, and if the feedback is still in phase.

Oscillator feedback circuits are arranged so that only one frequency of oscillation is obtained. This can be done by ensuring either:

- that the feedback is in phase at only one frequency; or
- that the amplifier gain exceeds feedback loop attenuation at one frequency only; or
- that the amplifier switches off entirely between timed conducting periods.

Oscillator circuits are of two types. Sine wave oscillators use the first two methods above for ensuring constant frequency operation. *Aperiodic* (or un-tuned) oscillators, such as multivibrators, make use of the third method. Oscillators are equivalent to amplifiers which provide their own inputs. They also convert the d.c. energy from the power supply into a.c.

Many oscillator circuits operate in Class C conditions. Even if the transistors start off with some bias current flowing, the action of the oscillator will turn off the bias for quite a large proportion of the complete waveform, making the transistor operate in Class C once it is oscillating. The reason for this will be clearer as you read through this chapter; it arises because a tuned circuit connected to an oscillator will continue to oscillate for a short time even when the transistor is no longer conducting. Aperiodic oscillators all operate in Class C.

Question 12.1

An oscillator circuit has had a transistor replaced and no longer oscillates. What is the most likely reason?

Sine wave oscillators

A sine wave oscillator consists of an amplifier, a positive feedback loop, and a tuned circuit which ensures that oscillation occurs at a single definite frequency. In addition there must be some method of stabilizing the amplitude of the oscillations so that the oscillation neither stops, nor builds up to such an amplitude that the wave becomes distorted by reason of bottoming or cut-off.

The most common types of sine wave oscillator are those which operate at radio frequencies, such as are used in the local oscillators for superhet receivers which use LC tuned circuits to determine the oscillating frequency.

- At this point you should revise the section on oscillators in the Level-2 book.

Hartley oscillator

Like most oscillator circuits, the Hartley oscillator exists in several forms, but the circuit illustrated here is a much-used type. The tuned circuit $L_1 C_2$ has its coil tapped to feed a fraction of the output signal back through C_4 to the emitter of Tr_1. Since an output at the emitter is always in phase with the output at the collector this feedback signal is positive. The base voltage of Tr_1 is fixed by the values of the resistors R_1 and R_2 with C_1 acting as an a.c. by-pass capacitor.

The amplitude of the oscillation is limited at the emitter because the transistor will cut off if emitter voltage rises to a value more than about 0.5 V below base voltage. The distortion of the wave-shape caused by this limiting effect is smoothed out by the flywheel effect of the tuned (or *tank*) circuit L1C2, which produces a sine wave voltage at the resonant frequency even when the current waveform is not a perfect sine wave.

Irrespective of how its feedback is arranged, the Hartley oscillator can always be recognized by its use of a tapped coil. Its frequency of oscillation, as is the case with all oscillators using LC tuned circuits, is given by the formula:

$$f = \frac{1}{2\pi\sqrt{LC}}$$

Example: What is the oscillating frequency of a Hartley oscillator which has a 1 5 µH inductor and a 680 pF capacitor in its tuned circuit?

Solution: Substitute the data in the equation to get:

$$f = \frac{1}{2\pi\sqrt{15 \times 10^{-6} \times 680 \times 10^{-12}}} = 1.58 \times 10^6 \text{ Hz, or } 1.6 \text{ MHz approx.}$$

Faults which can cause failure in an oscillator of this type include the following:

- bias failure caused by breakdown of either R_1, R_2, or R_3,
- a faulty by-pass capacitor,
- a leaky or o/c coupling capacitor C_4, or
- faults in either C_2 or L_1.

C_2 should be of the silver-mica type of capacitor. Some ceramic capacitors will not permit oscillation because of *lossy action* which means that the capacitor dissipates too much power to permit the circuit to resonate properly.

Colpitts oscillator

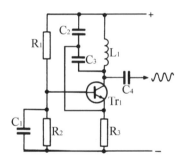

The example of this circuit shown here demonstrates its basic similarity to the Hartley oscillator. Instead of using a tapped coil, however, the Colpitts oscillator employs the combination of C_2 and C_3 to tap off a fraction of the output voltage to feed back into the base. This latter is biased and by-passed in the same way as in the Hartley circuit.

The same remarks about circuit operation, and about the several possible circuit configurations apply to the Colpitts as to the Hartley oscillator, but the formula for determining the frequency of oscillation is slightly different. Because the capacitors C_2 and C_3 are in series, it is the series combination C' (in which $1/C' = 1/C_2 + 1/C_3$) which tunes L_1 to give the output frequency. The formula must therefore use the value of C' rather than C_2 or C_3.

Question 12.2

Why must silver-mica capacitors be used in RF oscillator LC resonating circuits?

Activity 12.1

Construct the Colpitts oscillator shown in the diagram with the following component values: $C_1 = 0.1$ µF, $C_2 = 0.01$ µF, $R_1 = 10$K, $R_2 = 1$K5, $R_3 = 1$K, $C_3 = 0.001$ µF. The inductor L_1, should consist of 50 turns of 28-gauge enamelled copper wire close wound on a 10 mm diameter former fitted with a ferrite core. (Alternatively, the coil can be wound directly on a ferrite rod of the same diameter.)

Check the circuit and connect to a 12 V power supply. Connect the collector of the transistor to the Y-input of an oscilloscope, and link the negative line to the oscilloscope earth. Switch on, and adjust the oscilloscope to show the waveform from the oscillator. Measure the amplitude and frequency of the output wave.

Now observe the effects on the amplitude and frequency of oscillation of the following changes:

(a) Increasing the supply voltage to 15 V.

(b) Connecting an additional 0.001 μF capacitor in parallel with L_1.

(c) Reducing bias by connecting a 2K2 resistor in parallel with R_2.

(d) Increasing bias by connecting a 22K resistor in parallel with R_1.

(e) Reducing feedback by connecting a 0.1 μF capacitor in parallel with C_2.

(f) Increasing feedback by replacing C_2 by a 470 pF capacitor.

(g) Reducing the value of R_3 to 100R.

Note that changes (e) and (f) will both inevitably affect the frequency (if oscillation continues) because the capacitance of the tuned circuit has been changed in value.

- Note that the oscilloscope is the most certain way of checking that an oscillator circuit is working, and also of measuring the frequency. Another check is to measure the current drawn by an oscillating circuit, which will be much larger than normal if the circuit is not oscillating (because the oscillator operates in Class C when oscillating, requiring less bias current).

Tuned-load oscillators

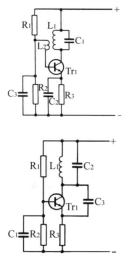

Oscillators of this type have a tuned circuit as the load of the transistor and use another component for feedback. The tuned-collector feedback oscillator shown in the diagram uses a feedback winding placed physically close to the winding of L_1 to extract a fraction of the output signal to invert its phase, and to feed it back to the base.

Remember that the output of a single-transistor common-emitter amplifier is always phase-inverted, so that another inversion must be carried out if feedback is to be taken from the collector to the base. Amplitude limitation is carried out in this circuit by the bottoming and cut-off action, and the output is smoothed into a sine wave by the resonant oscillations of the tuned circuit.

Failure of this type of oscillator circuit can be caused by the by-pass capacitor C_2 going o/c as well as by any of the biasing faults which can cause the Hartley oscillator to fail.

Another type of tuned-load oscillator is shown, left. This circuit uses a capacitor of very small value, C_3, to feed back part of the signal from the collector to the emitter. This type of oscillator is commonly used at high frequencies as a local oscillator in TV or FM receivers.

Crystal oscillators

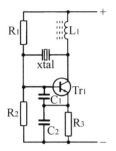

The use of a quartz crystal in place of a tuned circuit in an oscillator gives much greater stability of frequency than can be achieved in the same conditions by any LC circuit.

There is a great variety of crystal oscillator circuits, some of which use the crystal as if it were a series LC circuit, with others using it as part of a parallel LC circuit in which the crystal replaces the inductor. The example of a crystal oscillator shown here is a form of Colpitts oscillator but with the crystal providing the frequency-determining feedback path and some of the 180° phase shift between collector and base. The choke L_1 acts as the collector load across which an output signal is developed and also contributes to the phase shift.

In normal use, crystal oscillators are extremely reliable, but excessive signal current flowing through the crystal can cause it to break down and fail. The usual comments concerning bias and decoupling components apply here also. Oscillators of the same type and performance can also make use of surface acoustic wave (SAW) filters to determine frequency.

Note that all the circuits for sine wave oscillators can be constructed with MOSFETs in place of bipolar transistors. The use of MOSFETs has several advantages, particularly where crystal oscillators are concerned, because no input current is required at the gate of a MOSFET.

RC or divider-chain oscillators

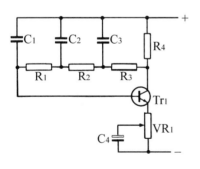

Oscillators that are required to operate at low frequencies cannot use LC tuned circuits because of the large size of inductor that would be needed. An alternative construction is the RC oscillator, and the diagram shows the basic outline of one type known as the *phase-shift oscillator*, or *divider-chain oscillator*.

Every RC potential divider must attenuate and shift the phase of the signal at the collector of the transistor. If the total phase shift at a given frequency is 180°, the signal fed back to the base will be in phase with the signal at the collector and the circuit will oscillate.

The output waveform will **not**, however be a sine wave unless the gain of the amplifier can be so controlled that it is only just enough to sustain the oscillation. All RC oscillators therefore require an amplitude-stabilizing circuit, which is usually provided by a negative feedback network. This network usually includes a component such as a thermistor whose resistance decreases as the voltage across it increases.

In this way, an increase of signal amplitude causes an increase in the amount of negative feedback which in turn causes the amplifier gain to decrease, so correcting the amplitude of oscillation. Failure in this feedback network will either stop oscillation, or cause severe distortion of the output waveform.

Activity 12.2

Construct the phase-shift oscillator of the diagram, using the following values:

R_1, R_2, R_3	220K
R_4	4K7
VR_1	1K
C_1, C_2, C_3	1 nF
C_4	100 μF, 15 V
Tr_1	2N3056, BC107, BFY50 or similar general purpose NPN type.

Adjust the potentiometer so that the circuit is just oscillating as detected by an oscilloscope connected to the collector of the transistor. Note that low-gain transistors may require the use of lower values for R_1, R_2, R_3.

With the circuit oscillating, note the d.c. voltage levels at the emitter base and collector of Tr_1. Sketch two cycles of the output waveform. Note the peak-to-peak output voltage, and periodic time of the output, and calculate the frequency. Note how the wave-shape changes as VR_1 is adjusted.

Repeat the d.c. voltage measurement when the circuit is not oscillating (adjust VR_1 to stop oscillation).

What is the effect of replacing one of the resistors R_1, R_2, or R_3 with a 47Ω value?

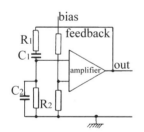

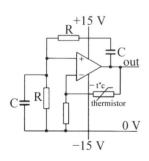

Another type of RC oscillator is shown here in an outline in which only the components which determine frequency are labelled. This type of oscillator is known as the Wien Bridge. It will be noted that the circuit uses feedback to the non-inverting (in-phase) input of the amplifier.

These RC circuits also require amplitude-stabilization if they are to produce sine waves of good quality, but they are capable of providing very low distortion figures (of the order of only 0.01 per cent) by good design. An outline example of a Wien Bridge oscillator circuit with provision for amplitude stabilization by a thermistor is illustrated here. The amplifier itself is a 741 IC. A practical circuit would need to contain some setting to allow for adjustment of the feedback so that the thermistor could hold the amount of feedback at a level that would only just permit oscillation.

A distorted output from a RC oscillator, or zero output, is nearly always caused by failure of a component in the amplitude-stabilizing circuit. Lack of output can also be due to a sudden loss of gain, such as could be caused by bias failure or by the failure of a decoupling capacitor.

Question 12.3

Some older phase-shift and Wein bridge oscillator circuits used a filament lamp in the stabilizing circuit. Can you suggest why?

Aperiodic oscillators

Revise at this point Unit 5 Outcome 2 of the Level-2 work, covering multivibrators. **Revise** the use of the 555 timer IC.

The 555 timer is an IC that is widely used as an oscillator or as a generator of time delays. The timing depends on the addition of external components (a resistor and a capacitor), and one very considerable advantage of using the IC is that the time delays or waveforms that it produces are well stabilized against changes in the d.c. supply voltage. As with other ICs, we shall ignore the internal circuitry (other than a block diagram) and concentrate on what the chip does, Figure 12.1.

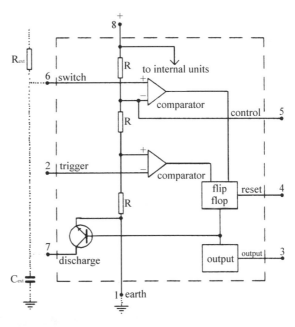

Figure 12.1 The internal block for the 555 timer and its pin connections

The chip pin connections for the usual 8-pin DIL form of the chip are shown also in Figure 12.1, using pins 1 and 8 for earth and supply positive respectively. The output is from pin 3, and the other pins are used to determine the action of the chip. Of these, pins 2, 6 and 7 are particularly important. Pin 7 provides a discharge current for a capacitor that is used for timing, and pin 6 is a switch input that will switch over the output of the circuit as its voltage level changes. For most uses of the chip, these pins are connected to a CR circuit whose charge and discharge determines the time delay or the wavetime of the output.

An astable pulse generator circuit is illustrated in Figure 12.2. Pin 6 is used to discharge the capacitor C_1 and is connected to the triggering pin 2. This ensures that the circuit will trigger over to discharge the capacitor when the voltage level reaches two-thirds of supply voltage, and will trigger back to allow the capacitor to charge when the voltage level reaches one-third of supply voltage. The frequency of the output can be adjusted by using a 100K variable in place of R_2 and, if needed, by using switched values of capacitance C_1.

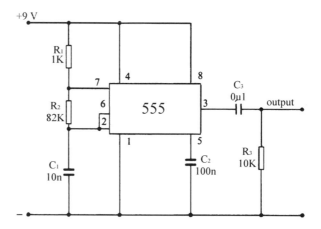

Figure 12.2   A 555 used in an astable pulse generator circuit

For the astable circuit illustrated, the waveform timings are given approximately by the following formulae:

Charge time	$0.7 (R_1 + R_2)C_1$
Discharge time	$0.7 R_2C_1$
Periodic time	$0.7 (R_1 + 2R_2)C_1$
Frequency	$1.44/(R_1 + 2R_2)C_1$

Activity 12.3

Construct an astable oscillator using the 555 timer with the component values shown in Figure 12.2. Use the oscilloscope to view the output waveform and measure its frequency. Sketch this waveform, indicating peak-to-peak voltage and periodic time. View and sketch also the waveform at pin 6, showing the voltage levels and the time for each part of the waveform. Measure the current taken by the circuit.

VCO

The voltage-controlled oscillator (VCO) is a circuit that is basically an oscillator whose output frequency can be controlled by an input (current or voltage) at one terminal. One way of implementing this is to use an astable circuit in which the rate of charge and discharge is controlled by a transistor rather than a resistor. For higher frequencies, the VCO is usually implemented by using a varactor as part of the tuned circuit of an oscillator. At one time, such reactance amplifier circuits were constructed using discrete transistors but the circuit is nowadays available only as part of an IC, usually a phase-locked loop

Phase-locked loop (PLL)

The phase-locked loop (PLL) is an important circuit which has been much more widely used since it first became available in the form of an IC (e.g. the National Semiconductor Co. NE567).

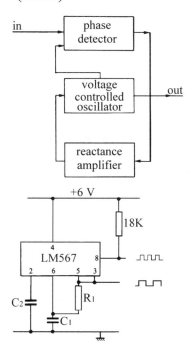

The block diagram of a PLL is shown here. The heart of the circuit is a VCO controlled by a reactance amplifier. The oscillator output is compared with that of an input signal. If this input signal is within a certain range (the *capture band*) of the oscillator frequency, control of the oscillator is taken over by the d.c. voltage generated by the comparator (or phase detector).

The reactance amplifier is a transistor stage whose action is made to resemble that of a reactance (current 90° out of phase with voltage) by positive feedback through a reactor. The circuit action as a whole will be familiar to the TV engineer as being that of the flywheel sync circuit. The versatility of the circuit is increased by the fact that a d.c. correction signal can be obtained from it, in addition to the corrected oscillator signal.

The PLL can be used simply as a precision oscillator whose frequency is set by the values of a single resistor and capacitor as in the example, left. In this application, the oscillator frequency is stable to within 60 parts per million (which means 60 Hz in 1 MHz), provided that a stabilized supply voltage is used. The value of C_2 is made ten times that of C_1, and the output frequency is given by the equation:

$$f_0 = \frac{1}{R_1 C_1}$$

where R_1 is in ohms and C_1 in farads.

Another mode of employment is as a very versatile detector and filter. The internal oscillator frequency can be set to practically any frequency, within wide limits, and the bandwidth over which the internal oscillator will lock to the frequency of the incoming signal can be set independently between limits of 0 per cent and 14 per cent of the frequency of the oscillator. With the bandwidth control set to minimum, the oscillator will lock very precisely to a frequency selected by the timing components.

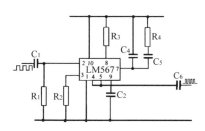

For example, if the input to the PLL is a crystal-controlled square wave, the timing components (R_3 and C_2 in the illustration) can be set so that the output frequency is at one of the harmonics of the frequency of the crystal (provided, of course, that this does not exceed the frequency range of the PLL). Every harmonic selected in this way will be perfectly locked in phase to the crystal frequency.

Such an arrangement is used in frequency-synthesizer circuits, since it can combine the versatility of the variable-frequency LC oscillator with the stability of crystal control. In addition, the components C_4, C_5, R_4 in the circuit illustrated form a filter circuit. Component values for this function can be calculated from the manufacturers' data sheet.

A major advantage of PLL circuits is that they make it possible to regenerate a signal which is practically lost in noise. Provided the input signal amplitude is enough to drive the PLL (only about 20 mV is needed for the NE567), any input frequency which is within the capture range of the frequency set by the timing capacitor and resistor will be locked in, and the output will be a waveform of that frequency which is free from both noise and interference.

A frequency-modulated signal can be demodulated using a PLL circuit, even if the signal is varying considerably in amplitude, because a change in the frequency of the input signal causes a change in the steady voltage of the phase detector signal. Since the voltage thus fluctuates exactly according to the frequency variations, it becomes the demodulated signal itself.

Activity 12.4

Examine the data sheet for a PLL such as the type 567. If a suitable IC is available, construct a test circuit as follows.

Select values for the timing components so as to give an oscillator frequency of 10 kHz, and for the bandwidth capacitor to give a bandwidth of 1 kHz. Follow the procedure laid down in the data sheet to select a value for the output filter capacitor.

Using an oscilloscope to monitor the output, feed to the input a signal of about 50 mV (r.m.s.) from a signal generator. Vary the frequency of the signal generator from 5 kHz to 15 kHz, and observe the frequency and amplitude of the output signal.

Sawtooth circuits

The ability of the integrator to convert a square wave into a wave with sloping sides makes it the natural basis for circuits used to generate the sawtooth waveform which produces timebases.

A simple timebase circuit is shown in Figure 12.3, following. While Tr_1 is conducting, the voltage at its collector is very low and C_1 is discharged. A negative-going pulse arriving (from an asymmetric astable multivibrator for example) at the base of Tr_1 will cut off Tr_1, and C_1 starts to charge through R_2. This integrating action generates a slow-rising waveform which forms the sweep part of the sawtooth.

As the trailing edge of the pulse reaches its base, Tr_1 is switched on again. C_1 rapidly discharges through Tr_1 and causes the rapid flyback at the end of the sweep waveform. Tr_2 is an emitter-follower acting as a buffer stage to prevent the waveform across C_1 from being affected by the input resistance of circuits to which the sawtooth is coupled.

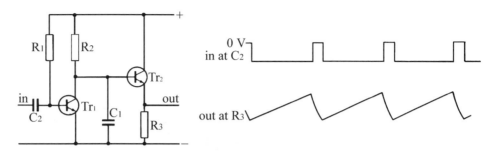

Figure 12.3 A simple timebase or sweep circuit, with waveforms

In such a simple circuit, the waveform across the capacitor will in practice be approximately a straight line if C_1 is allowed to charge only to a small fraction of the supply voltage. Thereafter the waveform will tend to bend over towards the horizontal. To prevent this, it is important to make the time-constant R_2C_1 much longer than the duration of the square pulse applied to the base of Tr_1.

Sawtooth-generator circuits normally use considerably more elaborate circuits than that shown in Figure 12.3, with the object of ensuring that the sweep voltage remains linear. One type of sawtooth generator uses constant-current circuits to replace R_2; another, the *Miller timebase*, uses negative feedback to keep the sweep waveform more truly linear.

The *Miller integrator* is a circuit based on the application of negative feedback through a capacitor. In the circuit of Figure 12.4, C_1 (the Miller capacitor) is connected between the collector and the base of a transistor. A rise of positive voltage at the input large enough to start the transistor conducting would, if C_1 were not there, cause a large negative change in voltage at the collector. Negative feedback through C_1, however prevents the voltage at the base of Tr_1 from rising faster than the rate at which C_1 can charge through R_1.

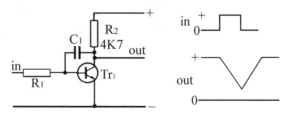

Figure 12.4 The basic Miller integrator circuit with idealized waveforms

This charging rate does not follow the usual exponential shape (while the transistor is conducting) because of the feedback through C_1, and a good linear sweep waveform is obtained. The principal application of the Miller integrator is in timebase circuits for oscilloscopes. If the gain of the transistor is high, the sweep waveform can be made very linear indeed.

A complete Miller timebase circuit for an oscilloscope is made considerably more elaborate than the basic circuit shown by the addition of other circuits whose purposes are (a) to discharge the Miller capacitor more rapidly (so obtaining rapid sweep flyback), and (b) to switch the input voltage at the times required to generate the timebase waveform.

Question 12.4

The collector to base stray capacitance of a transistor is often termed the Miller capacitance. Can you suggest why?

Activity 12.5

Construct the simple Miller integrator of Figure 12.4 giving R_1 a value of 1K and C_1 a value of 1 μF. Use a signal generator set to inject into the circuit square waves at 500 Hz, 1.5 V peak-to-peak, and observe the voltage waveform at the collector with the aid of an oscilloscope. Making no other changes, try the effects of the following set of values:

(a) $R_1 = 10K; C_1 = 1\ \mu F$

(b) $R_1 = 1K; C_1 = 0.1\ \mu F.$

Note that arrangement (a) gives C_1 a time constant ten times what it was before, and that arrangement (b) gives it a time constant which is one tenth of the previous value. Note the effects of these changes, and sketch the waveforms.

Answers to questions

12.1   The transistor has a lower gain so that oscillation is not possible.

12.2   Mica capacitors have very low loss; some other types are so lossy that they cannot be used in an oscillator circuit.

12.3   A filament lamp is non-ohmic and has a resistance value that increases as the current through it increases.

12.4   The capacitor provides negative feedback in the transistor that decreases its gain at high frequencies.

Unit 4

Outcomes

1. Demonstrate an understanding of logic families and their use in sequential logic circuits and fault finding on prepared circuit boards.
2. Demonstrate an understanding of time division multiplex (TDM).
3. Demonstrate an understanding of sequential logic.
4. Demonstrate an appreciation of common fault types and an understanding of fault finding procedures and test equipment.

13 Logic families and terminology

Device selection

Digital logic devices may be members of families of standardized types, or they may be tailor-made for a particular application. At one time, digital circuitry was composed almost entirely of recognizable chips, making replacement easy, but the trend in the last fifteen years has been to ASICs (application-specific integrated circuits) that are made specifically to carry out a large number of tasks in a circuit, and which are not generally available. Once the ASICs for a particular circuit are no longer available from the manufacturer, the repair of equipment that used the ASIC is no longer economically possible.

Despite the large-scale adoption of ASICs, a large number of digital chips are still used that are manufactured in huge numbers and from several suppliers, and it is important to be able to recognize and replace these devices. The four important features of a chip that is one of a standard family are:

- its type number,
- the type of packaging,
- the pin numbering and allocation,
- the type of output.

Preliminaries

Digital logic devices need to be provided with regulated supply voltages. Almost universally now, a single positive regulated supply is used, with an earth (zero voltage) return, but a few chips may need both regulated positive and negative supplies in addition to the earth connection. The supply current can range from a few nanoamps for some types of CMOS chips to several milliamps for the older STTL types.

All digital devices specify an acceptable range for input and output signal voltage. Because of the development of logic chip types starting with TTL (see later) it has been customary to design other types of digital circuitry to use the same levels (TTL levels). These are:

High level: +2.2 V to +5.5 V
Low level: 0 V to +0.8 V

Threshold level is another important term that relates to the changeover from level 0 to 1 or from level 1 to 0. If we consider an IC that operates with TTL signals whose nominal voltage levels are 0 V and +5 V, the signals that are acceptable at an input need not be *exactly* at these levels, they will normally conform to the high and low range illustrated above. The values of +0.8 V and +2 V are the threshold levels, and any input between these levels is indeterminate – you cannot predict what the output will be for an input that is in the threshold range. Digital circuitry must therefore be designed so that these intermediate levels exist only for a very short time during the change from 0 to 1 or 1 to 0.

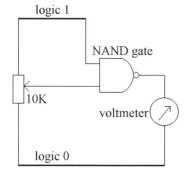

Activity 13.1

Connect a logic gate as illustrated in the diagram, using an STTL 7400 chip initially. Monitoring the output level, gradually raise the DC input level from zero to find the point where the output level just changes over. Note this voltage. Starting now from logic level 1, gradually reduce the steady input voltage until the chip switches over again, and note this voltage level also. Draw a diagram to show the threshold levels. Repeat with chips 74LS00, the CMOS 4011 and 74HC00.

The *noise margin* of a digital circuit is the amplitude of unwanted input pulse that will just make the device switch over. This refers mainly to an input at the 0 level. For example, if the maximum possible input voltage level for logic level 0 is 0.8 V, and the gate will actually switch when the voltage level reaches 1.4 V (a typical figure), then a positive-going noise pulse of 0.6 V peak amplitude will be enough to cause the gate to switch over. This value of 0.6 V is the noise margin (see Figure 13.1). It is more accurately described as the noise margin for logic 0, but since the noise margin for logic 1 is usually larger, this is the value that is normally quoted.

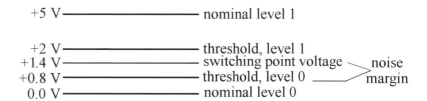

Figure 13.1 Threshold levels and noise margin

Note that noise margins are improved if digital circuits can be operated from higher voltages, so that the typical input levels of 0 V and +15 V that are used by the 4000-series of CMOS chips permit a much larger noise margin. The usual TTL type of circuits, however, are restricted to a +5 V supply, and noise problems are reduced by using a well-stabilized power supply with decoupling at each chip, and by using low-impedance inputs.

Modern microprocessor (and allied) chips often use MOS circuitry with a 1.45 V core supply used for processing actions, but using a 3.3 V level for (high) output. This reduces the dissipation within the microprocessor, but maintains a reasonably large noise margin at the output.

The input current of a device will be quoted at both signal levels. For TTL families, the input current at level 1 is very low, but the input current for logic 0 is substantial, several milliamps in the older designs. The modern logic chips have very low input current figures of the order of 10 pA (10^{-12} A) for either level of input. The output current range is usually 1 mA or more.

Rise and fall times are the times needed to change input voltage levels from one threshold voltage level to the other, and are of the order of ns or ps. A more important figure is *propagation delay*, which is the time between changing the input level and obtaining a corresponding change at the output. These times are of the order of a few ns for most of the standard families of chips, but times measured in ps are required for some purposes, notable fast A–D conversion.

Power dissipation for a chip can be high, several mW, for bipolar types, but is low, several μW, for MOS types. For many digital circuits, the power dissipation of the logic chips is of little importance because so many circuits use a microprocessor (dissipating 40 W or more) and memory chips (also with high dissipation), so that the cooling requirements are fixed by the needs of these other chips. Nevertheless, a packed logic board that uses only normal standard chip families may dissipate enough heat to require fan-assisted cooling.

Finally, the *absolute maximum ratings* for a logic chip define the levels that must not be exceeded, even for short periods. Any breakdown of regulation of a power supply will cause the absolute maximum voltage supply rating to be exceeded, so that power supplies must contain some form of over-voltage cutoff that will operate in the event of failure of regulation. Another common hazard is excessive input, because an input voltage might be obtained from some external circuit. MOS chips are protected by diodes to guard against damage from electrostatic voltages, but these protection diodes are not capable of passing the more substantial currents that would pass in the event of a failure of power regulation in a unit that fed signal into a logic circuit.

Source and sink, fan-out

Two terms that are very important to the operation of digital circuits are *source* and *sink*, applied to the current levels at an input or output of a digital IC. When we say that an output can source 15 mA, for example, we mean that the output can provide this maximum amount of current to whatever is connected to it. The connections will either be the inputs of other digital ICs, discrete circuits, or other loads such as micro-motors.

The ability to *sink* current at an output means being able to allow inward current flow without an unacceptable change in voltage level. If we say that the input of an IC requires a sink of 1 mA, for example, we mean that 1 mA can flow *out* of the *input*, and this current will flow to earth through the output terminal of the IC that drives this input, without causing the voltage at that point to rise unacceptably.

This leads to the term *fan-out*. The fan-out of a digital IC means the number of standard gate inputs that can be reliably driven from the output. For example, if a gate output can sink or supply 16 mA, and its input requires sinking of 1.6 mA, then one output can be connected to as many as ten inputs without compromising the ability of the circuit to operate.

Input connection is important for logic devices. No input must ever be left floating (disconnected) because this can lead to unexpected output changes, particularly at high clock rates when stray coupling can produce input voltages. The old STTL chips could be operated at slow clock rates, with an input left o/c, because the input to an emitter automatically ensured that the input level was at logic 1, but later types, and MOS chips in particular, must *always* have all inputs connected either to the output of another chip, or to one or other logic level through a resistor.

Question 13.1

If a chip circuit can provide a current of 1.6 mA at its output and requires to sink a current of 200 μA at an input, what is its fan-out value?

Activity 13.2

Using a two-input AND or OR gate, connect a two-way switch to one input and leave the other floating. Use a meter to measure the output level from the gate. Connect the switch so that one position provides a logic 1 input and the other provides a logic 0 input. From the output level, deduce whether the floating input is at 0 or 1 level. Repeat the measurements for several family types, both bipolar and MOS.

Device numbering

The two main logic families of devices are TTL and MOS, but there are several important subdivisions in each group. TTL is an abbreviation for *transistor–transistor logic*, and it originally meant a circuit in which each input was to the emitter of an IC transistor, with an output from a collector. A TTL circuit is designed to operate with a +5 V stabilized supply. Any voltage level above about 2.4 V at the input will be taken as being at logic level 1, and any voltage less than about 0.8 V at the input will be taken as being at logic level 0.

For example, Figure 13.2 shows a circuit diagram that approximates to the internal circuitry of a TTL gate of the original 74 family, whose type numbers start at 7400 and continue into five and six-figure numbers. The input stage consists of a transistor which has been formed with two (or more) emitters. This is comparatively simple to carry out when an IC is being formed, and 15 or more emitters can be formed on a single base of a transistor. The output stage makes use of the familiar *totem-pole* type of circuit, and the stages between these two will implement the gate action, and they need not concern us here.

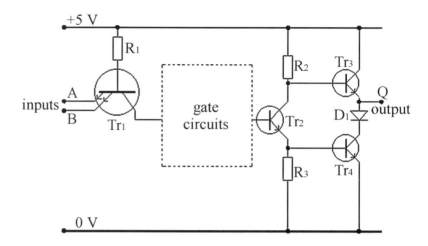

Figure 13.2   Typical input and output circuitry for STTL devices

This type of IC family is now termed STTL (S meaning standard), and is almost obsolete. The input impedance is low, so that an input current of about –1.6 mA must pass to an input in order to make the input transistor conduct. The output stage can sink or source a current of up to 16 mA, so that the maximum fanout is 10.

Bearing these quantities in mind, we can now look at the operation of the typical STTL circuit in detail. Imagine that the collector of Tr_1 is connected directly to the base of Tr_2. When both inputs, A and B are connected to logic 1, ideally +5 V, then no current will flow between the base of Tr_1 and either emitter, since the base voltage is not at a level sufficient to pass current (0.6 V above the emitter voltage). With no current passing between the base and either emitter, the collector voltage of Tr_1 will be high, with current flowing from the base of Tr_1 to the collector of Tr_1 and so into the conducting base of Tr_2. In this condition, both Tr_2 and Tr_4 will be conducting, and the output voltage at Q will be low because Tr_4 is conducting and Tr_3 is cut-off.

Imagine now that either one of the emitters is connected to logic 0. Current will flow through the base of Tr_1 to the emitter which has been connected to logic 0, and because this current is large enough to saturate the transistor, the collector voltage will be low, and no current will flow into the base of Tr_2, which will now be cut off. In this condition current flows through R_2 into the base of Tr_3, causing this transistor to conduct and thus connecting the output Q to the +5 V line. Tr_4 is cut off, and the output Q is at about 3.8 V, logic level 1.

Since the output at Q is low only when both inputs are at logic 1, the action is that of a NAND-gate, and the use of Tr_4 to connect the output to logic 0 ensures that comparatively large currents can pass from the terminal Q to the logic 0 line without raising the voltage level of the output above the guaranteed level 0 figure. Remember that when a transistor is saturated, its collector–emitter voltage is low, typically 0.2 V, and does not rise appreciably when current flows – Ohm's law is not obeyed by a transistor junction because the internal resistance changes as current changes.

Standard TTL chips carry a guarantee that a current of up to 16 mA can be sunk at the output when Tr_4 is conducting. Because the current that Tr_3 can pass is limited by the value of R_2 within the IC, there is generally no guarantee that this amount of current can be sourced because of the wide tolerance on the values of IC resistors, and designers are not advised to use circuits that require a source of 16 mA of current at Q, current which will pass through Tr_3.

The input currents flow only to the emitters that are connected to logic level 0, and the value of resistor R_1 will pass a current of 1.6 mA maximum. Since the maximum current (guaranteed) which can be sunk at an output is 16 mA, ten times the maximum current that must be sunk at the input to hold the input level at a guaranteed logic 0 level, fixing the fan-out of a circuit of this type at 10.

All members of the 74 family of STTL chips use inputs that are transistor emitters, so that they pass current at logic level 0 and no current at logic level 1. The propagation delay, which is the time between changing the level at the input and finding a change at the output is in the range of 11 ns to 22 ns ($1ns = 10^{-9}$ seconds). The power dissipated per gate under average switching conditions is 10 mW, and the typical operating frequency is around 35 MHz.

Low-power Schottky (LS)

At the time of writing, STTL chips are manufactured only for replacement purposes, though millions of STTL chips are still in use. Another type of TTL circuit, the low-power Schottky (LS) has in the past replaced the standard variety because of the twin advantages of high speed of operation and lower power dissipation. The Schottky chips are identified by the use of 74LS in the type numbers, so that type 7400 is an STTL AND-gate, but 74LS00 is a low-power Schottky AND-gate.

Low propagation delays that are obtained in STTL by using large currents inevitably lead to larger chip dissipation, because the current is flowing through the integrated components which are all part of the IC. The LS range of TTL ICs avoids this difficulty by using a different principle, relying on the use of Schottky diodes, components which can be made easily in integrated circuit form.

The Schottky diode, whose symbol is illustrated, left, is formed by evaporating aluminium on to silicon, and its remarkable feature is its very low forward voltage when it is conducting, of the order of 0.3 V. This feature is used in two ways. One application is to carry out the logic action using the diodes in circuits similar to those used in the early DTL (diode transistor logic) ICs.

The other use is in preventing the transistors in the circuit from saturating. A transistor is saturated when it is fully conducting, with the base passing more current than is needed to make the collector circuit conduct fully, since the collector current is limited by the value of collector load resistor. In this state, the collector-to-emitter voltage will be low, about 0.2 V, as compared to the 0.6 V which will exist between the base and the collector.

When a transistor is saturated, there is a comparatively large amount of slow-moving charge in the base layer, and when the transistor is switched off by connecting the base terminal to the voltage level of the emitter, this stored charge will permit current to flow between the collector and the emitter for a time which can be as long as a microsecond, until the charge is neutralized. This restricts the speed at which a switching circuit can be operated, but if the transistors in switching circuits are not allowed to saturate, a very considerable increase in switching speed is possible. This cannot, however, be achieved by normal biasing methods, particularly within an IC. Another advantage of using Schottky junctions is that they operate using majority carriers, so that the storage effect does not apply.

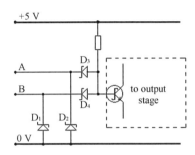

The diagram, left, shows the typical input circuit of a NAND-gate of the 74LS family which uses Schottky diodes for logic, and also uses diodes of the same construction connected between the collector and base of each transistor to prevent saturation.

The Schottky connection within the transistor is between the base and the collector so that when the transistor collector voltage approaches the saturation value, current from the base circuit is diverted by the Schottky diode into the collector circuit. This prevents the base current from reaching the value which would cause saturation. The Schottky symbol is shown on the transistor base to make it clear that the Schottky diode structure exists within the transistor. The transistor circuit which is used within these Schottky TTL ICs is designed to make use of current stabilisers in addition to other methods of preventing saturation

The 74LS series operates at the same voltage levels as the STTL 74L series, but the maximum output current is 8 mA, and the input current at logic level 0 is –0.4 mA, one quarter of the STTL current, so that the fan-out is 20. The propagation delay is in the range 9 ns to 15 ns, appreciably lower than that of STTL, and the average power per gate is about 2 mW. The typical switching frequency is 40 MHz.

Open-collector gates

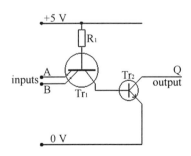

Though most of the gates of the 74 and 74LS series of TTL digital ICs make use of the totem-pole type of output circuit which has been illustrated, a few gates use what are called *open-collector* outputs. This means that the phase-splitter and one half of the output stage is missing, so that the output terminal of such a gate is simply the collector terminal of a transistor. If this gate is to be used in normal logic circuit applications, a load resistor, connected between the output and the +5 V supply line, must be added externally.

When the transistor, of which this resistor is a load, switches off, the voltage at the collector will rise. This rise of voltage is comparatively slow, however, because the stray capacitance at the collector is being charged through the load resistor rather than through the very much lower resistance of a transistor, as it would be if the totem-pole circuit were used. The rise time of a pulse when stray capacitance is being charged through a

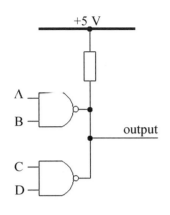

CMOS gate circuits

load resistor is therefore noticeably slower than the fall time when the stray capacitance is discharged by the conducting transistor.

The advantage of using open-collector outputs is that they can be used to make a *wired-OR* connection, which is impossible when the totem-pole type of output stage is used. A wired-OR connection is obtained when the outputs of two open-collector gates are connected together, and also to a load resistor.

In this example, the output is logic 0 when either (or both) gate output are at logic 0, so that the gate action is (A NAND B) OR (C NAND D). If this connection were made using normal gate ICs with totem-pole outputs then it would lead to the destruction of the gates when any one gate had inputs which caused a 0 output and the other had inputs that caused a 1 output. Gates of the usual totem-pole type must never have their outputs directly connected. Wired-OR connections are seldom used nowadays.

NOTE: Before you read this section, you should revise the principles of MOSFETs, see your Level 2 text.

By far the most common semiconductor technology in current use is that of metal-oxide-semiconductor field-effect transistors (MOSFETs). Three types of ICs can be manufactured using P-channel and N-channel FETs. PMOS ICs use P-channel FETs exclusively, NMOS ICs use N-channel FETs exclusively, and CMOS (C meaning complementary) ICs make use of both P- and N-channel FETs in a single circuit. PMOS methods were initially used for manufacturing microprocessors and similar chips, but were superseded by NMOS. Fast versions of CMOS are used in lap-top computers.

One family of CMOS devices uses the 4xxx type of number, illustrated later, but the most common types currently in use follow the 74xx type of numbering, with lettering to distinguish the types, such as 74HC, 74HCT, and 74AHC.

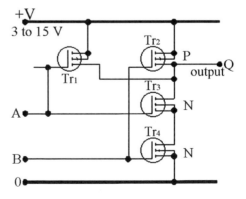

Figure 13.3 A typical CMOS circuit, in this example a NAND gate

A typical CMOS circuit, that of one NAND-gate of the CD4011A quad NAND IC, is shown in Figure 13.3. In this circuit, Tr_1 and Tr_2 are both P-channel types; Tr_3 and Tr_4 are N-channel types. The P-channel FETs will be switched into conduction by a logic 0 input at their gates, since their sources are connected to the positive supply.

The N-channel FETs will be switched into conduction by a logic 1 input at their gates, because their sources are connected to the 0 voltage line. With either or both gate inputs at logic 0, the P-channel FETs will conduct, keeping the output high. Only when both inputs are high can both N-channel FETs conduct, and thus connect the output to the logic 0 level. The action is therefore that of a NAND-gate. A NOR gate can be created using the same set of components by connecting the N-channel FETs in parallel and the P-channel FETs in series.

These CMOS ICs can operate with a wide range of supply voltages, typically 3 V to 18 V, and with very small currents flowing, typically 5 µA. The logic 0 and 1 voltages are normally very much closer to the supply voltage levels than is possible with older bipolar designs of the TTL, ECL or I^2L types. For example, using a +5 V supply, a logic 1 voltage of +4.95 V and a logic 0 voltage of 0.05 V can be obtained. This makes for much better noise margins than can be achieved with either STTL or LSTTL devices.

The input current is always negligibly small because the inputs are connected to MOSFET gates, and the output currents are typically about 0.5 mA maximum. The fan-out figure for low-frequency operations can be very large, 100 or more, but the value decreases as the frequency of operation is increased. This is because the small currents that are available at the output must be capable of charging and discharging the capacitance at the input of each gate which is connected to the output. This requirement for charging and discharging stray capacitances also increases the total dissipation of the IC as the frequency of operation is increased.

A simple gate, for example, which has a dissipation of 1 µW at a frequency of switching of 1 kHz, may have a dissipation of 0.1 mW at an operating frequency of 1 MHz. This factor limits the operating speed of the earlier types of CMOS circuits, and leads to these types (the 4000 series) being used in low-speed applications rather than for high-speed machine-control or computing applications. They are widely used where speed is not of primary importance, though.

- The very high insulation resistance of the gates makes them very susceptible to damage from electrostatic charges, and modern CMOS ICs are manufactured with a network of diodes connected to the inputs which will conduct whenever the voltage between gate and source or gate and drain becomes excessive. These diodes will protect for static voltages of up to 4 kV, but if higher voltages are likely to be encountered stringent earthing precautions must be taken. For example, operators may be required to use metal wrist straps that are earthed, and work on a conducting earthed surface. The safest way to work with CMOS devices is to earth all pins together until they are inserted into place and connected.

- Note that walking along a nylon carpet can generate voltage levels in excess of 16 kV.

CMOS families originally used numbering from 4000 upwards, and the numbers were later supplemented with the letters BE (older types) and UBE. These letterings mean, respectively, buffered and unbuffered. Later, new families of CMOS used the numbering of TTL devices, with the letters HC, HCT or AHC, so that a 74HC00 was a gate equivalent to the 7400 or 74LS00 in action, but with greatly reduced power consumption. These CMOS equivalents can be used, subject to some caution on their characteristics, as replacements for older bipolar STTL devices.

The older 4000BE series have typical propagation delays of 125 ns to 250 ns, power dissipation per gate of 0.6 µW, and a typical switching frequency of 5 MHz. The UBE family feature propagation delays of 90 ns to 180 ns and slightly higher typical frequency ratings. The 74HC family are direct replacements for 74LS types, with propagation times of 8 ns to 15 ns, power dissipation of 1 µW, and typical frequency of 40 MHz. The 74HCT is very similar, but with slightly longer propagation delays. The most recent CMOS family is designated 74AHC, with propagation delays of, typically, 5 ns, power dissipation of 1 µW, and a typical frequency of 100 MHz.

The lettering used for more recent 74-series logic chip families can be confusing, though the most common types have already been listed. Table 13.1 shows a more comprehensive list of known types in the year 2001. The quest for higher speed and lower dissipation continues, however, and several of the types in this list may become obsolete, with new types being introduced, during the lifetime of this book.

Activity 13.3

Connect up a NAND or NOR gate with one input connected to earth. Connect a multimeter to measure the output logic level, and provide an input from a two-way switch so that the input can be connected to either logic 0 or logic 1. Investigate the effect of inserting a resistor between the switch and the gate input, using values of 100R, 1K, and 10K.

With no resistor in the input circuit, check the effect on output voltage of resistors connected (a) to logic 1, (b) to logic 0. Try resistors of 4K7, 47K and 470K.

Question 13.2

Why can you not quote a single value of fan-out for a CMOS circuit?

Table 13.1 Some TTL families available in 2001

Lettering	Description	Lettering	Description
74F	Fast Logic	74AS	Advanced Schottky Logic
74ABT	Advanced BiCMOS Technology	74AVC	Advanced Very-Low-Voltage CMOS Logic
74AC	Advanced CMOS Logic	74HC	High-Speed CMOS Logic
74AHC	Advanced High-Speed CMOS	74LS	Low-Power Schottky Logic
74ALB	Advanced Low-Voltage BiCMOS	74LV	Low-Voltage CMOS Technology
74ALS	Advanced Low-Power Schottky Logic	74LVC	Low Voltage CMOS Technology
74ALVC	Advanced Low-Voltage CMOS Technology	74LVT	Low Voltage CMOS Technology
74ALBT	Advanced Low-Voltage BiCMOS Technology	74S	Schottky Logic Speed-Power Product

Table 13.2 shows a comparison of the most common families of devices that use either the 7400 or the 4000 type numbers.

Table 13.2 IC families summarized

	STTL	LSTTL	CMOS	74HC	74AC
V+supply	5 V	5 V	3–15 V	5 V	5 V
Imax/1	40 µA	20 µA	10 pA	10 pA	10 pA
Imax/0	−1.6 mA	−0.4 mA	10 pA	10 pA	10 pA
Imax/out	16 mA	8 mA	1 mA*	4 mA	4 mA
Delay	11–22 ns	9–15 ns	40–250 ns*	10 ns*	5 ns
Power	10 mW	2 mW	0.6 µW	1µ W	1 µW
Frequency	35 MHz	40 MHz	5 MHz	40 MHz	100 MHz

NOTES:

V+supply = normal positive supply voltage level

Imax/1 = maximum input current for logic level 1

Imax/0 = maximum input current for logic level 0 (continued opposite)

Imax/out = maximum output current

Delay = propagation delay in nanoseconds

Power = No-signal Power dissipation per gate in mW or µW

Frequency = Typical operating frequency

$1pA = 10^{-12}A$

* These quantities depend on the supply voltage level.

Question 13.3

Name the logic families corresponding to type numbers 4001, 74LS240, 74HC10, 7400.

Tri-state logic chips

Some circuits, notably computer boards, feature chips that are described as *tri-state*. This does not mean that they use three logic states, only that the chip can be isolated from inputs and output by not applying an enabling pulse. The purpose of this is to allow a set of chips to be permanently connected to the same set of input and output lines, selecting which chips are in use by their enabling inputs. Tri-state logic chips are therefore extensively used in any circuits that make use of bus lines to connect different portions of the circuitry.

Packaging

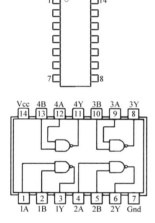

The packaging of the standard families of logic gate circuits is conventionally as DIL (dual-in-line) 14-pin units. The single in-line (SIL) type of packaging is now used mainly for items such as thick-film resistors and capacitors rather than for active circuits, and the use of surface mounting components (SMC) is normally reserved for custom (ASIC) chips.

Because the 14-pin package is more than is needed for a single gate, most combinational logic chips are packaged as a set, typically of four gates (or six inverters). Of the fourteen pins, two are reserved for earth (pin 7) and V+(pin 14). The conventional way to show pin allocations is a diagram that shows the top view of the chip with logic gate symbols and connections drawn in as illustrated here.

Answers to questions

13.1   8.

13.2   CMOS fan-out figure depends on frequency of clock signals.

13.3   CMOS, low-power Schottky, high-speed CMOS, standard TTL.

14 Digital communications

PCM and sampling

If the data that is to be communicated has originated in a computer, perhaps as text in ASCII code, then there is no need to have any conversion system for communication since the data is already digital. The majority of communications, however, still originate as sound, mainly voice (as in telephone communications), and such signals must be converted to some digital form before they enter the communications links. These communications may be by copper wire, by fibre-optic cable, or by modulated radio waves using either terrestrial or satellite broadcasting.

Activity 14.1

Connect two digital circuits by a fibre-optic link and verify that signals are being passed.

We class digital signals by their structure and information transfer rate. For convenience, binary digits (*bits*) are organized into various sized groups. For example, 4 bits form a *nibble* that can represent one hexadecimal character, 8 bits form a *byte* or *octet* which can be used to represent an alpha-numeric or control character from the 7-bit based ASCII code set (the extra bit is commonly used for parity check, see later). A *word* may consist of several bytes, often 2, 4, 8 or more long. The data transfer rate may then be quoted in bits, bytes or characters per second.

Alternatively, the term *Baud rate* may be encountered and this can cause confusion. The Baud rate for a communications systems means the fundamental frequency of the carrier wave, and by using suitable modulation techniques, it is possible that each transmitted symbol can represent more than 1 bit of information. If each symbol represents 4 bits, then the bit rate is four times the Baud or symbol rate. Only if each symbol represents 1 bit are the bit and Baud rates equal. See later for phase shift modulation methods that have a bit rate considerably higher than the Baud rate.

The term *Baud rate* is often used loosely and incorrectly (particularly in computing) to mean bits per second. For this reason, the term *symbol rate* is now more often used in digital TV engineering applications.

The form of digital signal that is universally employed for data is PCM, pulse code modulation, meaning that each unit of data is represented by a binary number. The ASCII code for text uses one byte of binary code for each text character, but data in analogue form, such as sound or picture data, has to be converted. Such conversions make use of sampling and coding.

Sampling and coding

Sampling means finding and storing the amplitude of a signal. By using sampling, the analogue signal is broken down into a set of voltage levels, each of which can be stored until it is converted (coded) into a binary number that is proportional to the amplitude of the sample. The output of these sampling and coding stages will be the PCM signal that can then be transmitted along digital links. The important features of this process are the rate of sampling and the number of binary digits used for each coding action.

The advantages of using this system are:

1. It allows digital devices to be used throughout the transmission network.

2. It can provide a significantly higher transmission speed than can usually be achieved with analogue processing.

3. It provides for improved transmission quality when electrical noise is present. Any noise component can be reduced using signal regenerators and error detection/correction techniques.

4. It is more compatible with the digital switching techniques used to control distribution and is a natural technique to use for systems involving an optical fibre link.

5. Encryption/decryption can easily be added for data security.

6. Where necessary, signal compression/bit rate reduction techniques can be employed to minimize the bandwidth requirement.

7. For many applications, time division multiplexing (TDM) can be used more effectively than frequency division multiplex (FDM) that is common for analogue transmission systems.

8. For systems involving reception and retransmission, signal regeneration can be used at the intermediate stage to improve signal quality.

TDM and packets

Time division multiplexing (TDM) is a method for making efficient use of a communications channel (which may be a radio frequency band, a copper cable or a fibre optic link) so as to carry more than one channel of information. The principle is that a set of bytes from one channel is transmitted, followed by a set from another channel, and so on, until the process is repeated with bytes from the first channel again. This is possible only if

1. the bytes belonging to each channel can be identified at the receiver,

2. the bytes for any one channel can be assembled in memory at the receiver as fast as they are required,

3. the rate of sending bytes can be controlled so that the memory at the receiver does not overflow.

Schemes for TDM all depend on the use of *packets* of data. A data packet is a set of bytes of channel information, such as PCM from a sampled source, together with bytes that indicate the start and end of the packet, with identification to ensure that a packet intended for one channel cannot be confused with a packet for another channel.

Figure 14.1 shows a typical packet structure, in this example, the packet structure for use with the ISDN service operated along conventional copper telephone lines. The *Start* flag indicates the beginning of the packet and can contain a sequence of bits designed to aid synchronization. This is followed by an address that identifies the package. For ISDN, this contains ID codes for the originator of the data and the destination of the data. The control section then contains the packet number so as to ensure that the packets are assembled into the correct order at the receiving end.

The word *flag* is used to mean a bit or set of bits that are not part of the data but are used as markers.

start	address	control	user data	frame check	end
(a)	(b)	(c)	(d)	(e)	(f)

Figure 14.1   The ISDN packet structure

The data bytes are then sent. One very commonly used convention is to use 256 bytes in each packet. The data bytes are followed by a frame check that can be used to provide error protection and correction, and then an end flag to signify the last bit in the packet.

At the receiving end, logic circuits identify each packet and assign it to its correct channel. In each channel, the packets are assembled in memory and fed to their destination, which might be another storage device (such as a hard disk) or another converter such as a D–A converter that will convert the digital data stream into analogue format.

- NOTE that this system requires some form of control of transmission rate, so that when the memory that is used for packet assembly (the buffer) cannot be over-filled. The usual scheme is that the rate at which the buffer assembles information is equal to the rate at which D–A conversion is carried out, and the number of channels being carried is calculated to supply the correct rate of packets in each channel. More complex methods are needed when the rate of reading packets might vary.

Activity 14.2

Use a TDM system to transmit several channels of data through a single path, and separate the channels again.

Question 14.1

A digital package is often compared to a postal package. List two main points of similarity.

Cable and fibre links

Digital data, encoded and multiplexed as packets, can be transmitted over fixed links, such as copper or fibre-optic cables, or by way of modulated radio waves. In general, some form of modulation of a carrier is used no matter how the data is transmitted, because modulation systems can be devised that allow very efficient use of the bandwidth of the transmission system (bit rate much higher than Baud or symbol rate). The earlier systems used for digital modulation were amplitude shift keying (ASK) and frequency shift keying (FSK).

Amplitude shift keying methods modulate the amplitude of the carrier according to the discrete level of the data signal. For binary modulation, the carrier is simply switched ON or OFF to represent 1 or 0 respectively. This technique is also known as on-off keying (OOK). Because ASK and noise signals have similar characteristics (a change of signal level), this form of modulation is not used nowadays because of its poor S/N ratio and hence a relatively high bit error rate (BER).

Binary frequency shift keying involves switching the carrier wave between two set frequencies. One form of this, much used for small computers that stored data on audio-tape, is known as Kansas City modulation, for which bursts of eight cycles of 2400 Hz or four cycles of 1200 Hz are used to represent 1 or 0 respectively. Other frequency values can be used and these may or may not have an exact 2:1 ratio. Small differences are often introduced so that beat notes between the two frequencies will not introduce bit errors. FSK has a better BER performance than ASK under the same conditions, but it requires a wider transmission bandwidth.

Phase shift keying (PSK) is a single frequency method in which the data signal is used to switch the carrier phase. Typically for binary transmission a 0 produces no effect whilst a 1 generates 180° of carrier phase shift. Of these three methods, PSK has the best BER performance and the narrowest transmission bandwidth, so that modern methods of digital communications all make use of some form of phase-shift keying.

For cable transmissions, either copper or fibre-optic, multi-phase PSK, in which the modulated carrier can carry a number of different phase shifts, can be used to advantage. Figure 14.2(a) shows an example of 8-PSK where each vector can be used to represent 3 bits of information so that the Baud rate is just 1/3 of the bit rate. The concept of 4-PSK or quadrature PSK (QPSK) can be usefully extended in several ways. For example, if each of the four vectors are permitted to have any one of four different amplitudes, then each vector can be used to represent 4 bits. Figure 14.2(b) shows one example of such quadrature amplitude modulation (16-QAM) in which the signal points in the matrix form a constellation. Cable digital TV systems can make use of even more bits of PSK, and one common system is 64-QAM.

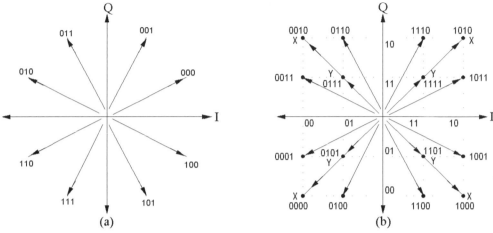

Figure 14.2 Digital modulation: (a) 8-PSK, (b) 16-QAM

Question 14.2

For a 16-QAM system, what is the ratio of bit rate to Baud rate?

Satellite transmission

The main problem for any transmission of digital data by satellite is the very feeble signal that is received on Earth. This is small enough on the dish size that is used for analogue reception, and is smaller still on the tiny digital dish. The main problem, then, for satellite reception, is the poor signal-to-noise ratio, which can be overcome only if the digital signals are modulated in a way that is very resistant to noise corruption. The compensating advantage is that the microwave frequencies used for satellite transmission allow the use of very large bandwidths, typically 30 MHz or more.

The modulation system that is used is termed QPSK (quadrature phase-shift keying), and, as the name suggests, each of two phases can be modulated with one bit, so that using the two phases (at 90° to each other), allows two bits to be carried on each carrier wave or set of carrier waves. This type of modulation is much less likely to suffer corruption by noise than any system that codes a greater number of bits per wave. The current microwave band for digital satellite broadcasting is in the range around 11.7 to 12.1 GHz.

At the receiving end, the signal is picked up by a small microwave dish, and converted to a lower frequency, typically to a frequency in the range 950 MHz to 1.75 GHz, by a mixer stage, the *low-noise block* (LNB), located at the dish. This avoids further losses that would be incurred leading a microwave frequency signal down a long cable. The local oscillator in the LNB is set to a fixed 9750 MHz, so that the selection of a carrier frequency is made at the receiver rather than at the dish. Power to the LNB is provided from the receiver or set-top box through the connecting cable. The LNB uses special low-noise transistors made from gallium arsenide rather than from silicon, and the layout of the components is very important because even a few millimetres of conductor can have an appreciable inductive reactance at microwave frequencies.

- The frequency of the local oscillator in the LNB is usually set by a ceramic resonator (altering capacitance) with a screw adjustment. This is placed close to a portion of the circuit but is not electrically connected other than by stray coupling.

At the set-top box, this intermediate frequency is converted to a QPSK digital signal, and then through a set of stages that reverse the coding steps that were done at the transmitter end, eventually being converted to analogue video and audio signals.

Terrestrial

Terrestrial broadcasting uses the existing land-based transmitters to broadcast a signal directly to existing aerials. If the receiving aerial is capable of receiving an analogue signal of good quality, then it should be well up to the task of receiving the digital signals on the same range of UHF carrier frequencies. The difference is that digital signals for the UK are not modulated directly on to one single carrier frequency. Once again, we in the UK have chosen a system that is different from that used in the USA, so making our TV receivers more expensive (as we did by selecting PAL rather than NTSC).

The system that has been chosen, and which is used in the digital receivers that are now available (at very high prices) and in the set-top boxes for the ITV-Digital (formerly called *On-Digital*) transmissions, is called *coded orthogonal frequency division multiplexing*, COFDM. The carrier is coded as if it consisted of a set of 1705 separate carriers, each of 4 kHz spacing, and each modulated, using QAM-64, with part of the digital signal. This makes the transmission rather like a parallel cable with 1705 strands, each strand carrying 6-bit signals.

- The ICs that code and decode this type of signal are very complex. There was some doubt at the design stage if the task could be done, and it was for this reason that the 1705 figure was adopted in place of a proposed 8000 carrier system. In fact, the 8000 carrier system could have been used because the technology of manufacturing ICs has progressed just as rapidly as that of the systems using them.

The use of COFDM allows a lower signal to noise ratio to be tolerated, so that digital transmitters can work on lower power without sacrificing the service area. For example, my local digital transmitter (Sudbury) uses power outputs of between 7 and 8 kW (effective radiated power) for BBC and ITV, using channel 49 for BBC and 68 for ITV (and C4). The digital transmitters for the same location work at 7.5 kW for channel 39, but at only 1.5 kW for channel 54 and 1.1 kW for channel 50.

At the receiver or set-top box, the COFDM signal is reassembled into a continuous stream of digital bits, and is then decoded for program selection in the same way as is used for satellite or cable reception.

Secondary encoding

With all forms of signal transmission, it is important to minimize the effects of noise which is a destroyer of information. With analogue systems, the effect is measured by the signal to noise (S/N) voltage, current or power ratio. There is no simple way of measuring S/N for digital signals, and the quantity that **can** be measured is the energy/bit per watt of noise power (E/B).

For television transmission systems the bit error rate (BER) is in common use. This is simply the ratio of the number of incorrect (error) bits to the total number of bits transmitted per second. If the noise power becomes comparable with the energy in each bit, bit errors are produced. Thus the degradation of the digital S/N ratio leads to a bit error rate (BER), and this BER is used as a measure of degradation of a digital signal.

The use of sampling and PCM produces a stream of binary digits from an analogue source, and though this form of digital data can be used within a computer, it is not ideal for either recording or transmission. For example, recording systems cannot cope with a long stream of bits of the same type because these are rather like a very low-frequency analogue signal. Most recording systems are equally upset by very frequent changes of level, such as 0101010101..., and methods, classed as secondary encoding, have to be used to ensure acceptable recording on magnetic tape, magnetic disc or compact disc (including DVD).

Similar problems are encountered when we try to transmit PCM signals. Digital communications systems are more robust than the equivalent analogue systems in the presence of noise. This is clearly seen by reference to Figure 14.3 which shows how a clean signal (b) can be regenerated simply by slicing the received signal (a) at appropriate levels. Figure 14.3(c) further compares the two types of signals under the same degree of decreasing S/N ratio at the input.

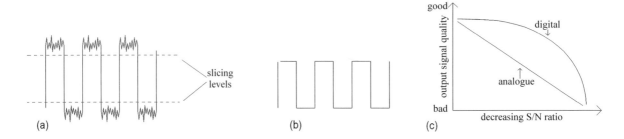

Figure 14.3   (a), (b) Regenerating a digital signal, (c) comparison of signal behaviour in noisy conditions

Whilst the analogue system degrades gracefully before it fails, the digital system is relatively unaffected by the noise until a point is reached, when the system suddenly crashes as the bit error rate (BER) rises. However, the advantages of the digital technology does not end here. There are a number of ways in which the information can be accurately recovered even from signals with a high BER, providing that the time scale is not critical.

Information recovery

At the simplest level of communication between two computers, a system known as Automatic ReQuest for repeat (ARQ) is available. Reference to the ASCII (American Code for Information Interchange) code table shows that two special codes are available, ACK and NAK. If a distant receiver detects a code pattern without errors, it transmits via a return channel the code ACK (Acknowledge). However, if errors have been detected, transmission of the NAK (Negative acknowledge) code automatically generates a request for a repeat transmission of the last block of signal code.

Furthermore, bit errors can be detected and even corrected in a digital system using a technique known as forward error control (FEC). This is so called because the means of error detection/correction is contained within the transmitted message stream. This is achieved by the addition of extra redundant bits which when suitably processed, are capable of identifying the errors. Either method makes extra demands upon the spectrum; FEC requires additional time or bandwidth to include the extra bits, whilst the ARQ system requires a free return channel.

The prime causes of bit errors are *white noise* and *impulsive noise*. The former produces errors that are completely uncorrelated and random in occurrence, whilst the latter creates a loss of bit stream synchronism which leads to bursts of errors.

There are three classes of error that need to be considered;

1. Detectable and correctable,
2. Detectable but not correctable, and
3. Undetectable and hence uncorrectable.

For any errors detected as type 2, the error can be concealed even if it cannot be corrected. The concealment options are:

a   Ignore the error and treat it as a zero level,

b   repeat the last known correct value, or

c   interpolate between two known correct values.

Question 14.3

ARQ is very effective for ensuring that communications between computers are 100 per cent accurate. Why can we not use such methods for digitized audio and TV transmissions?

Forward error control can significantly enhance the resistance to noise interference of a digital communications system in a noisy environment. Comparatively simple methods can be applied to coded text that uses the ASCII code system. This 7-bit code allows for $2^7 = 128$ different alphabetic, numeric and control characters. The most commonly used digital word length is 8 bits or 1 byte ($2^8 = 256$), therefore there is space for one extra redundant bit in each code pattern. Note that this spare bit is not available for extended ASCII codes (used by word-processing programs) that use all eight bits.

Even and odd parity

A single-error detection (SED) code of n binary digits is produced by placing n–1 information or message bits in the first n–1 bit positions of each word. The nth position is then filled with a 0 or 1 (the parity bit), chosen so that the entire code word contains an **even** number of 1s. For example, using an 8-bit byte to convey a 7-bit item of data allows the 8th bit to be used as a parity bit. If a code word that has been adjusted for even parity is received over a noisy link and is found to contain an odd number of 1s, then an error must have occurred.

Alternatively, a system might use odd parity, where the nth bit is such that the code word will contain an odd number of 1s. In either case, a parity check at the receiver will detect when an odd number of errors has occurred. The effects of suffering two (or any other even number) errors is self-cancelling, so that these will pass undetected. The even or odd parity bits can be generated or tested, using Exclusive OR or Exclusive NOR logic respectively.

Such a pattern of bits is described as an (n,k) code, n bits long and containing k bits of information. The simple example of ASCII code is an (8,7) code. It thus follows that there are $n - k = c$ parity (or *protection*) bits in each code word. The set of 2^k possible code words is described as a *block* or *linear* code. ASCII uses $2^7 = 128$ words in a block code. At one time, all memory used in PC computers allowed for 9-bit storage, so that a parity bit was available for checking memory integrity. This scheme is no longer used because memory is more reliable now, and there are better ways of checking memory integrity (such as by successive writing and reading methods).

The simple parity scheme that provides only a low level of error control, is suitable only for systems that operate in relatively low levels of noise. The weak point of the system is that a single error may affect a parity bit, causing a valid byte to be rejected. In addition, the system is useful only in a communications link where a fault detected at the receiver can be signalled to the transmitter so that a block can be sent again.

Question 14.4

Write down the versions of the following codes when even parity coding has been used: 0011010, 1001000, 1011000

Check-sum technique

This error control scheme is often used with magnetic storage media where data is stored in long addressable blocks. It is a simple scheme in which the digital sum of all the numbers in any block is stored at the end of each block. Recalculating the check sum and comparing the original and the new values after each data transfer, quickly tests if any read errors have occurred. For example, if the following binary words 0101 (5), 0011 (3), 1010 (10) and 0010 (2) are valid then the check sum would be 10100 (20). If after a read operation the numbers became 0101 (5), 0011 (3), 1001 (9) and 0010 (2), the check sum should be 10011 (19). Comparison of the check sums show that a read error has occurred and the block should be read again.

Weighted check-sum technique

The simple check-sum technique can only identify when an error occurs and cannot indicate the position of the actual error. However, this problem can be overcome by using a *weighting scheme* that employs a series of prime numbers (any number that cannot be divided perfectly by any other number except 1 and itself).

By multiplying each data number by a prime number, the checksum that is formed carries position information, and subtracting the true (transmitted) checksum from the false checksum (obtained by repeating the checksum formation on the received faulty data) will give the multiplier number for the data digit that is in error.

This can be most easily explained using a series of denary numbers. Suppose we use the prime numbers 1, 3, 5, and 7 and the decimal values to be stored and read are 5, 3, 9, and 2. The weighted check sum would be calculated by multiplying each data number by the corresponding prime number. In this example, it would be:

$$1 \times 5 + 3 \times 3 + 5 \times 9 + 7 \times 2 = 5 + 9 + 45 + 14 = 73.$$

The sequence would thus be stored as 5, 3, 9, 2, 73. If on reading this became 5, 3, 8, 2, 73, the recalculated check-sum would be:

$$1 \times 5 + 3 \times 3 + 5 \times 8 + 7 \times 2 = 5 + 9 + 40 + 14 = 68.$$

The check-sums are different so an error has occurred, but the difference is $73 - 68 = 5$. Therefore a unit error has occurred in the 5-weighted value, the data number that was multiplied by 5 in this example, which was originally 9 and has become 8 because of the error.

When this scheme is applied to a binary number, correction is simple, because if the digit 0 is an error, the true value is 1 and vice versa.

Hamming codes

Much more advanced error correcting codes have been devised, and one type has been named after R. W. Hamming, who was the originator of much of the early work on error control.

Figure 14.4 shows in simplified form how a message may be expanded with redundant check bits. These are usually interleaved with the message bits and placed in positions 2^0, 2^1, 2^2, etc. in the encoded pattern.

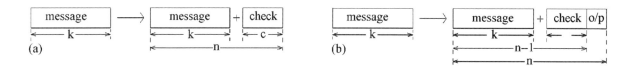

Figure 14.4   Hamming codes: (a) single error correcting (SEC), (b) single error correcting/double error detection (DED)

The mechanics for the encoding/decoding process can be explained using Table 14.1 for an example of a (7,4) block code (block length n = 7, message length k = 4 and parity bits c = 3).

The message to be transmitted is (0011) and these bits are loaded in the 3rd, 5th, 6th and 7th positions respectively. The three parity checks are carried out, to determine the values to be placed in positions 1, 2, and 4. In general, the xth parity check bit is given by the sum modulo-2 (the remainder after division by 2) of all the information bits in the positions where there is a 1 in the xth binary position number. The transmitted code word thus becomes 1000011. If this is now received over a noisy link as 1000001, there is an error in the 6th (110) position. The receiver decoder then performs the same three parity checks and generates the results shown, left.

1st check   0
2nd check   1
3rd check   1

The reverse of this series (110) which is called the *syndrome* (syndrome being a medical term for the symptoms of a disease) points to an error in bit 6 (binary 110 = denary 6) position. When the error position has been pin-pointed, it can be corrected by simply inverting bit 6 (in this example). An all-correct transmission would have produced an all-zero syndrome.

Table 14.1   Using Hamming codes, and finding a syndrome

Bit Position	1	2	3	4	5	6	7
(Binary)	001	010	011	100	101	110	111
Message/Parity	(P)	(P)	(M)	(P)	(M)	(M)	(M)
Check 1	*		*		*		*
Check 2		*	*			*	*
Check 3				*	*	*	*
Message			0		0	1	1
Parity bits	1	0		0			
Transmitted Code Word	1	0	0	0	0	1	1
Received Code Word	1	0	0	0	0	0	1
Recheck of Parity (reverse order)	0	1		1			

Syndrome = 110 = 6 (denary)

By adding an overall parity (O/P) check bit as shown in Figure 14.4, the single error correction capability is extended to double error detection. The error patterns indicating the following conditions:

1. No errors; zero syndrome and overall parity satisfied,

2. single correctable error; non-zero syndrome and overall parity fails,

3. double errors, non-correctable; non-zero syndrome and overall parity satisfied.

Hamming code schemes are particularly useful for systems that require a high level of data integrity and operate in noisy environments that create random bit errors.

Cyclic redundancy check (CRC)

This method of error detection is most effective in combating burst errors. To generate the code for transmission, three code words are used. These are a *message* code word k, a *generator* code word G (selected to produce the desired characteristics of block-length and error detection/correction capability), and a *parity check* code word c. As for block codes, the transmission code word length is given by n = k + c. During encoding, the message k is loaded into a shift register and then moved c bits to the left, to make room for c parity bits. The register contents are then divided by the generator code word to produce a remainder that forms the parity check code word c, which is then loaded into the remaining shift register cells.

Thus if the total code word were transmitted and received without error, this when divided by G, would yield a zero remainder. The last c bits can then be discarded to leave the original message code word. If however, an error occurs, then division by G leaves a remainder code word that acts as a syndrome. There is a one-to-one relationship between this and the error pattern, so that any correctable errors can be inverted by the error correcting logic within the decoder. The effectiveness of these codes depends largely on the generator code word which has to be carefully selected. A further advantage is that CRC can be operated with microprocessor based coding, when the system characteristics become re-programmable.

Other code error control techniques

There are a number of other techniques that have been developed from the basic Hamming concepts to deal with both random and bursts of errors. These include BCH codes for random error control, Golay codes for random and burst error control and Reed-Solomon codes for random and very long burst errors with an economy of parity bits. Reed-Solomon coding is used for the CD recording system.

Activity 14.3

Use a logic tutor to investigate a coding circuit and produce a table that confirms the information on the data sheet.

Primary codes and pulse shapes

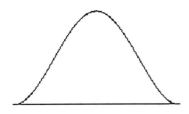

The bit error rate can be minimized by using pulses of maximum width and/or amplitude, the obvious choice being a square shape. However this introduces a number of problems. To pass a square wave, a transmission channel requires a wide bandwidth. To retain a good approximation to a square wave requires that the channel bandwidth should extend up to at least the 13th harmonic of the fundamental frequency. In any case the transmission of such pulses through a typical channel will produce *dispersion* or pulse spreading, which leads to a type of error called inter-symbol-interference and an increase in the bit error rate. Increased pulse width reduces the signalling speed and an increase in pulse amplitude introduces further problems. The final pulse shape is thus a compromise. One particularly useful pulse shape is described as a *raised cosine*. This shape (see left) is chosen because half of the pulse energy is contained within a bandwidth of half the bit rate.

Secondary codes and formats

A code format is a set of rules that defines clearly and with no room for mistakes the way in which binary digits can be used to represent alphabetic, numeric, graphic and control character symbols. The PCM code is one useful format, but other formats can be used with advantage to reduce the BER when noise is likely to corrupt a transmission.

Shannon's rule states that the communication channel capacity in bits/second, is related to the available bandwidth and the S/N ratio. It also shows that these two parameters can be balanced to maximize the channel capacity for an acceptable bit error rate suitable for a particular service.

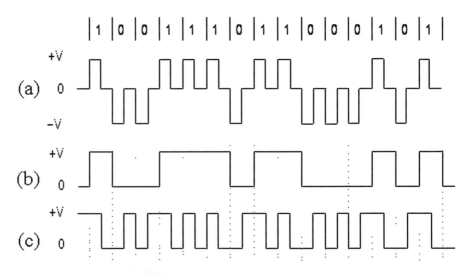

Figure 14.5   Commonly used code formats: (a) Return to zero, (b) NRZ version, (c) Manchester (bi-phase)

To take advantage of this trade off, binary code formats are re-designed by inserting extra bits into the data stream in a controlled way. Some of the formats used are shown in Figure 14.5. The general aim is to minimize the number of similar consecutive bits and balance the number of 1s and 0s in the message stream. The greater number of signal transitions is used to improve the locking of the receiver clock and so reduce bit errors. The balance between the number of 1s and 0s produces a signal without a d.c. component in its power spectrum. This allows a.c. coupling to be used in the receiver and reduces its low frequency response requirement. The commonly used codes are generated and decoded using dedicated integrated circuits (ASICs).

The return-to-zero (RZ) format where a 1 is represented by a half width pulse and a 0 by a negative pulse is shown in Figure 14.5(a). This is a simple way of coding PCM, and it was used extensively in early telegraphy, but it is seldom used now because the reduced-width pulses require more energy to be transmitted per bit than in other coding methods.

A non-return-to-zero (NRZ) basic code is one in which a 1 is signified by a full width pulse and a 0 by no pulse, shown in Figure 14.5(b), where a 1 is signified by a signal transition at the centre of the bit cell and a 0 by no transition. Further variants of the NRZ code which have the same characteristics include inversions of these two states.

The Manchester code shown in Figure 14.5(c) is a bi-phase format in which each bit in the original signal is represented by two bits in the derived format. The basic rule for the transform is that 0 is represented by 01 and 1 by 10. This ensures that there are never more than two identical bits in series. One variant of this format uses the opposite of this transform whilst two further variants adopt their inverses.

CD systems also use EFM (eleven-to-fourteen modulation) as a way of overcoming the problems of recording on CD media. In this system, the 8-bit code has been extended to 11 bits with parity bits, and each 11-bit number is transformed into a 14-bit number for recording. The 14-bit numbers are chosen so as to avoid the repetition of a digit or the rapid changes between 0 and 1 that can be troublesome in a recording or replay action.

Other codes

Code forms like NRZ and Manchester can be used on code that is to be recorded or transmitted, but for other purposes, other codes can be devised. The normal 8-4-2-1 binary code is by far the best for carrying out straightforward arithmetic, and is used almost exclusively in circuitry that is mainly concerned with arithmetic in which no conversion to denary is needed.

The 8-4-2-1 type of code is not the only binary code that we can use, however, and several others are likely to be encountered, depending on what type of applications you have for digital logic circuits. One very common variant of the 8-4-2-1 code is BCD, binary-coded decimal. This is a very popular form of coding when numbers have to be shown on LED or LCD displays, because each digit of such a display is one denary digit.

In a BCD system, then, each denary digit is represented by 4-bit binary code. Table 14.2 demonstrates what this implies. The conversion between BCD and denary is simple, and conversion to a form suitable for driving a display is also simple. The conversion between BCD and binary is, however, not quite so simple for numbers of more than one digit, and arithmetic with BCD is also not so simple. An added disadvantage is that BCD requires more storage space for any given number than 8-4-2-1 binary, though it is possible to devise systems in which number accuracy of floating point numbers can be much better at the cost of very much slower arithmetical processes.

Table 14.2   Denary and BCD equivalents

Denary	BCD	Denary	BCD
0	0000	5	0101
1	0001	6	0110
2	0010	7	0111
3	0011	8	1000
4	0100	9	1001

BCD is really just an adaptation of 8-4-2-1 binary, but Gray code is quite a different form. As Table 14.3 shows, a Gray code scale does not use columns to indicate the value of a bit, and you always have to use a table to convert a number in Gray code to a denary or a binary 8-4-2-1 number. The Gray code numbers are for 4-bits only, because Gray code is used either in BCD form, or in a scale of 16 (hexadecimal).

Table 14.3   Gray code and denary equivalents

Denary	Gray code	Denary	Gray code
0	0000	8	1100
1	0001	9	1101
2	0011	10	1111
3	0010	11	1110
4	0110	12	1010
5	0111	13	1011
6	0101	14	1001
7	0100	15	1000

The advantage of Gray code is that only one bit ever changes at a time during a count up. The change from 7 to 8, for example is from 0100 to 1100, rather than from 0111 to 1000 in the 8-4-2-1 binary code. Gray code has particular advantages for conversion of quantities like the rotation of a shaft into binary form, because if the shaft is in a position between the angles represented by numbers 7 and 8, then there may be reading errors caused by some binary digits that are changing between 0 and 1, and the fewer digits that change, the lower the chances of error.

There are other forms of binary code, such as Excess-3, which is a form of BCD in which 3 is added to each digit before coding into 8-4-2-1 binary. This means that the smallest value of code is 0011 and the largest is 1100, with anything below 0011 or above 1100 being an error. This makes error-detection easier, and has also the considerable advantage that BCD numbers coded in this way can be manipulated by the same circuits as ordinary binary. The Gray code and the various forms of 8-4-2-1 code are, however, by far the predominant methods of coding that you are likely to encounter.

Answers to questions

14.1   Each package carries an address and is to a single destination.

14.2   4 bits per symbol.

14.3   There is no two-way connection so that a request for repeat transmission is impossible. In addition, there may not be time for retransmission.

14.4   10011010, 01001000, 11011000

15

Sequential logic

R-S flip-flops

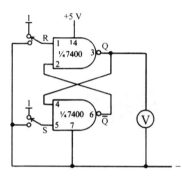

A flip-flop is a circuit of the bistable type. IC flip-flops tend to be of complex designs which would be uneconomical to manufacture with discrete components. The diagram shows a very simple flip-flop circuit, the R-S flip-flop which can be formed (among other ways) by using two NAND gates connected as shown. R-S flip-flops are also, of course, available as ICs in their own right, but they are very seldom used now for reasons we shall appreciate later.

The R-S flip-flop has two signal inputs, labelled respectively R (reset) and S (set), and two outputs, labelled Q and Q# or $\overline{Q}$, with the hash sign (#) or bar indicating an inverse in the usual way.

> The Q, Q# type of notation is preferred because this convention is easier to type using either a typewriter or a word-processor than the system that uses the overhead bar to represent the inverse.

The truth table for this circuit demonstrates an important point distinguishing this and all other flip-flop circuits from simple gate circuits. Flip-flops belong to a class of circuits called *sequential logic* circuits in which the output Q has a value that depends on the *previous* value of the inputs as well as on their present value. The output of a combinational logic gate, by contrast, depends only on the values of inputs which are present at the same time as the output(s).

Thus the truth table output for the inputs R = 1 and S = 1 depends on what values of R and S existed immediately before the R = 1, S = 1 state. If the previous state was R = 0, S = 1, then the arrival of R = 1, S = 1 gives Q = 1. If the previous state was R = 1, S = 0, then the arrival of R = 1, S = 1 gives Q = 0.

R	S	Q	Q#
0	1	1	0
1	1	1	0
1	0	0	1
1	1	0	1

- Note that there is no line R = 0, S = 0 for this form of R-S flip-flop. Such an input would give Q = 1, Q# = 1, which is not permissible. Moreover, R-S flip-flops can only be used when the state R = 0, S = 0 cannot possibly arise. For this reason other types of flip-flops, such as the D-type and the J-K type, are used in most logic applications.

Activity 15.1

Connect two gates of a 7400 in the form of an R-S flip-flop, as in the diagram above. Use voltmeters to indicate the states of Q and Q, and switches to provide the inputs. Draw up the truth table.

Question 15.1

What is the form of the truth table for an R-S flip flop constructed from two NOR gates?

Clocked flip-flops

The simple R-S flip-flop changes state when either of the inputs becomes, or is set to, zero. Most digital circuits require flip-flops that change state only when forced to do so by means of a pulse, called a *clock pulse*, arriving at a separate input. In a set of such *clocked* flip-flops, all the flip-flops in a circuit can be made to change state at the same time.

Clocking has the advantage that the inputs need only be set an instant before the arrival of the clock pulse, because changes of state at the inputs at any other time have no effect on the output. Between clock pulses, the output remains as it was when set by the last clock pulse.

D-type flip-flop

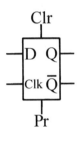

The D-type flip-flop shown here triggers on the leading edge of the clock pulse. The logic value on the D (data) input is then transferred to the Q output, thus delaying the data by the period of one clock pulse.

Table 15.1   D-type flip-flop truth table

Pr	Clr	Clk	D	Q	Q#
0	1	X	X	1	0
1	0	X	X	0	1
0	0	X	X	1	1
1	1	↑	1	1	0
1	1	↑	0	0	1
1	1	0	X	Q	Q#

Notes: X = either state, ↑ = goes high.

Reference to the truth table, Table 15.1, shows the action of the Preset (Pr) and Clear (Clr) inputs to set the initial output states of the device. When Pr = 0, Q is set to 1 and when Clr = 0, Q is set to 0. The states of both Clk and D are irrelevant (X) at this moment of time because Pr and Clr override them. Only when Pr = , Clr = 1 will the clock pulse transfer the data, and even then only as the clock pulse reaches a high value.

- The state where Pr = Clr = 0 represents an unstable condition in a D-type flip-flop, and is prohibited.

The D-type flip-flop is often referred to as a *latch*, because the output value at the Q terminal remains at its previous setting, regardless of the input at D, until a clock pulse is applied. A latch is a very valuable component of digital circuits because it performs a form of memory action. A sample-and-hold circuit for A-D conversion, for example, uses a latch to hold a sampled value until it can be processed.

Activity 15.2

Assemble a latching circuit using a D-type flip-flop and verify the action by drawing up a state table.

J-K flip-flop

The J-K flip-flop, which also contains steering logic to avoid the output problem of the R-S flip-flop at R = 0, S = 0, also has a greater range of output controls as shown by Table 15.2. This type of flip-flop has three signal inputs labelled S(set) and R(reset) (or sometimes pre-set and clear) which work independently of the clock pulses, and the usual two outputs Q and Q# (with Q# always the inverse of Q for any combination of inputs).

Table 15.2   State table for a J-K flip-flop

J	K	Q_t	Q_{t+1}	Comment	J	K	Q_t	Q_{t+1}	Comment
0	0	0	0	No change	1	0	0	1	Set to 1
0	0	1	1	No change	1	0	1	1	Set to 1
0	1	0	0	Reset to 0	1	1	0	1	Toggle action
0	1	1	0	Reset to 0	1	1	1	0	Toggle action

Q_t = state now; Q_{t+1} = state after next clock pulse

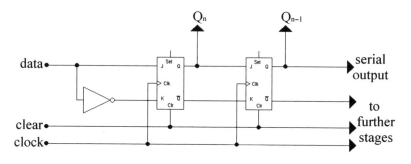

Figure 15.1   J-K shift register with parallel access

Figure 15.1 shows two sections of a J-K shift register for serial output but with parallel access. As can be seen from Table 15.2, the J and K inputs not only control the flip-flop action, but also provide the Set and Reset functions.

If the complementary outputs of the J-K device change state before the end of a clock pulse, then because of the internal feedback, the inputs will also change. This can cause the device to oscillate until the end of the pulse and leave the output in an indeterminate state.

To avoid this so-called *race around*, the clock pulse duration should be small compared with the propagation delay. For high speed operation, this is avoided by using a master/slave device as shown in Figure 15.2 where the NAND logic gates act as switches. The state table is the same as for the simple J-K device.

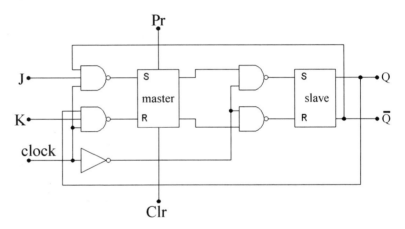

Figure 15.2 Master/slave J-K flip-flop

Data is transferred in two stages on both edges of the clock pulse. On the positive edge, the input gates are operative and allow the data to be loaded into the master flip-flop. At this time the output logic gates are open circuit. On the negative pulse edge, the switch states reverse, the master is isolated from the input and the data is transferred to the slave stage to provide the output. The master-slave design is now predominant, and it is most unlikely that you will find the older version in use.

The binary counting action occurs when both J and K inputs are taken to + 5 V (in the case of TTL circuits). Two clock pulses arriving at the clock terminal give rise to one pulse at the Q output, while a pulse of inverse polarity will simultaneously reach the Q# output.

Note, by the way, that clock pulse inputs to digital circuits must have short rise and fall times. The inputs to gates also should never be slow-changing waveforms. The reason is that the gain of these circuits, considered as amplifiers, is very large, so that a slow-changing waveform at the inputs can momentarily bias the circuit in a linear mode, so making oscillation possible. Oscillation also can cause mis-counting and erratic operation.

Activity 15.3

Connect the circuit shown in Figure 15.3 using a 7476 IC, which contains two master-slave J-K flip-flops, preferably constructed on a breadboard. Apply clock pulses either from a slow pulse generator or from a de-bounced switch (see below), and observe the output indicator LEDs.

Now connect the clock terminal (pin 6 of the second flip-flop in the package) to the Q-output (pin 15 of the first flip-flop), and re-apply the slow clock pulses. Observe the indicators.

Remember that no pin must ever be left disconnected.

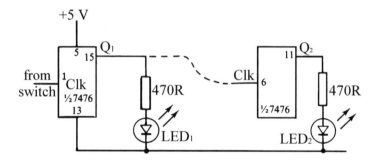

Figure 15.3   Circuit for Activity 15.3

Counting and dividing with bistables

A fundamental feature of all clocked bistable devices is that they can be used so as to divide the input number of clock pulses by two. Thus for every two clock-pulse inputs, the Q output changes once. This concept forms the basis of electronic binary counters, an example of which is shown in Figure 15.4.

The pulse sequence to be counted is the input to FF_1, whose Q output now provides the clock for the next stage, and so on. A count input to the first stage thus ripples through the circuit from one flip-flop to the next, which is why a counter of this type is often called a ripple, a ripple-through or an *asynchronous* counter. Note that the output count should be read from right to left, because the most significant bit (MSB) is, by convention, shown on the right of the drawing. A chain of n flip-flops will, connected in this way, produce a counter that divides the number of input pulses by a maximum of *n*.

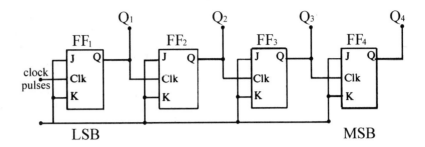

Figure 15.4   A ripple or asynchronous counter circuit

The ripple counter waveforms are shown in Figure 15.5. Since $J = K = 1$, the Q output toggles, or changes, on the edge of every negative-going waveform.

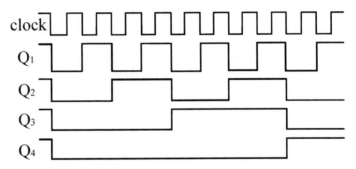

Figure 15.5   Ripple counter waveforms

- When the Q output of every stage of a flip-flop is used to provide the clock pulses, the circuit counts down from a preset number, to produce a ripple-down counter.

If the control circuit shown in Figure 15.6 is added between each flip-flop, it is possible to produce a counter which will count up or down according to the state of the control line. If control is logic 1, the circuit will count up; while if the control is logic 0, it will count down.

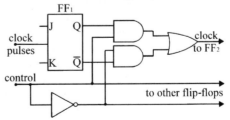

Figure 15.6   The up-down counter control circuit used between stages

Divide by N counters

You might need to divide by a number N which is not a power of 2; for example, to divide by 10 for a denary counter. A counter that carries out a division by N is called a *modulo-N counter*. The design method that is used for such counts is that of an *interrupted count*, meaning that the counter contains sufficient flip-flips for a longer count, but is interrupted after the count of N and reset to zero. To construct such a counter n flip-flops are needed, such that n is the smallest number for which $2^n > N$. A feedback circuit detects when a count of N has been reached so that the circuit can be reset to zero. For a decimal counter $n = 4$, because $2^4 = 16$ which is the nearest number greater than 10. Denary 10 is 1010 in binary, so that the feedback must detect this pattern to reset the counter to zero.

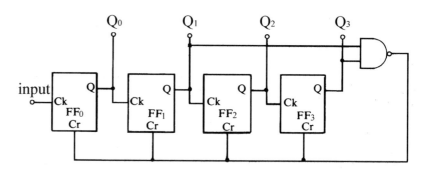

Figure 15.7   Asynchronous or ripple decade counter

A typical asynchronous or ripple decade counter is shown in Figure 15.7 with the least significant bit (LSB) on the left at FF_0. The output count appears on terminals Q_0 to Q_3. This output needs to be decoded to display the count in decimal form. As soon as $Q_1 = Q_3 = 1$, the Clear (Clr) inputs immediately reset all the Q outputs to zero to restart the count. Note that Q_1 and Q_3 first become 1 after the tenth pulse and then are quickly reset to zero. This generates a very narrow pulse at the output which requires additional circuitry to suppress it. Ideally Q_1 and Q_3 should go to zero immediately on the tenth pulse.

Synchronous counters

These ripple or asynchronous counters suffer from the serious disadvantage that they are relatively slow. The propagation delay is the time required for the counter to respond to an input pulse. In the worst case, when all the flip-flops are at logic 1, the next pulse must reset all the flip-flops to zero. This input pulse thus ripples through all the flip-flops in the circuit. If the total propagation delay is longer than the time between pulses, some pulses simply are not counted. If the operation is changed so that all the flip-flops are clocked at the same time (synchronously) the propagation delay is reduced to that of a single flip-flop thus increasing the maximum count rate that can be used. For a TTL 4-bit counter, the maximum count frequency might rise to above 50 MHz compared with about 25 MHz for a ripple counter. Integrated high-speed IC MOS counters can operate at much higher speeds.

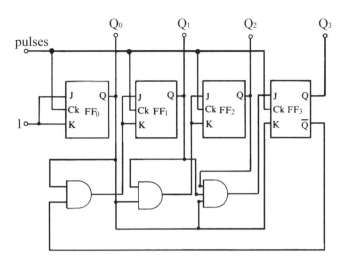

Figure 15.8 A synchronous decade counter

Synchronous counters can be constructed using J-K flip-flops and gates, and a circuit for a synchronous decade counter is shown in Figure 15.8. Since the bistables are all clocked in parallel, it works much faster than the ripple counter. The J-K flipflop changes state (toggles) only when $J = K = 1$ at the time of a clock pulse. The toggle action in this circuit is controlled via the Q outputs and the AND gates. Before any bistable can toggle, all the earlier bistable Qs must be at logic 1. Because $J_0 = K_0 = 1$, FF_0 toggles on every clock pulse. While $Q_3 = 0$, $Q\#_3 = 1$ and when $Q_0 = 1$, $J_1 = K_1 = 1$ and FF_1 toggles. In a similar way, FF_2 toggles when $Q_0 = Q_1 = Q_2 = 1$. When $Q_3 = 1$, $Q_3\# = 0$,so that $J_3 = 1$ and $K_3 = 0$ (because $Q0 = 0$ on the tenth pulse). At this time all the flip-flops reset to zero.

IC counters

An interrupted count to nine, followed by resetting, however it is achieved, is the basis of BCD counters, in which each counter unit (a set of four flip-flops) is used for one column of a denary number. BCD counting allows for the display of a count number in denary rather than in binary, so that BCD counters are preferred for any application in which a count is to be displayed rather than simply used.

BCD counters are available in IC form in the 74 series, and the simplest unit is the 7490. This is one of the very few IC counters that is classed as asynchronous, and is designed mainly for use with displays. The design is, in fact, a split counter with one scale-of-two (a single flip-flop) and one scale of five which uses three flip-flops connected as a synchronous counter. For decade counting, the output of one unit is connected to the clock input of the other, so that the overall action is asynchronous. Because of the asynchronous action, the Q outputs will not change simultaneously, and this can give rise to spikes on the outputs. These are unimportant if the output is used to operate a display, since the display does not store the spike output, but it makes a counter of this type unsuitable if the outputs are connected to latches.

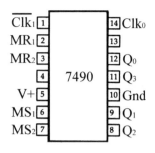

The pinout for the 7490 is shown here, left. The pins labelled MR_1 and MR_2 are reset inputs which must both be taken to logic 1 to reset the counter, since they are inputs to a NAND gate. The inputs MS_1 and MS_2 are similarly used to set the most significant and least significant flip-flops to 1 (setting the count at 1001, denary 9). Table 15.3 shows how the MR and MS inputs control the action.

Table 15.3   The use table for the 7490 type of ripple counter chip

MR1	MR2	MS1	MS2	Q0	Q1	Q2	Q3
1	1	0	X	0	0	0	0
1	1	X	0	0	0	0	0
X	X	1	1	1	0	0	1
0	X	0	X		counting		
X	0	X	0		counting		
0	X	X	0		counting		
X	0	0	X		counting		

The counter is clocked by the trailing edge of the clock pulse. The maximum ripple delay, from the clock pulse input to the Q_3 output, is of the order of 100 ns for standard TTL, but around 50 ns for the LS version of the chip.

For decade counting with synchronous counters we would normally make use of an integrated unit such as the 74160 type. This is not so simple as the 7490 unit that we considered earlier. To start with, this is a totally synchronous counter, and its internal flip-flops are triggered at the leading edge of the clock, with the outputs also changing at this time. The counter is fully presettable, meaning that the count number can be preset at any stage so that a count from, for example, 3 to 9 rather than 0 to 9 can be carried out if necessary. Only an up-count is available, and the counter allows a hold state in addition to its presetting state, making it fully programmable.

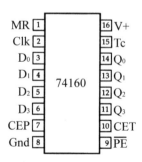

The diagram, left, shows the pinout for this unit. The MR input is a resetting input which is active when taken low and would normally be used shortly after power has been applied to the counter, before counting begins. This is an important point, because in any device that consists of a set of flip-flops, application of power will cause some flip-flops to set to a 1 output, other to reset to a 0 output. A counter cannot therefore be used immediately power has been applied; it must be reset before any counting pulses are applied. The reset action, as usual, is completely asynchronous and this is reflected in the table by the X which means 1 or 0 (the don't care state). The CP pin takes the clock pulse input, and the arrows in the table are a reminder that the changes take place on the leading (0 to 1) transition of the clock pulse, which should have a short rise time.

The CEP and CET inputs are normally high for counting, but taking either or both to level 0 will cause the count output to hold its existing state. This action must not be used while the clock pulse input is low, only after the leading edge of the clock pulse when the clock input has settled to a high state. The PE pin is used for parallel loading of the flip-flops, allowing a number to be preset before counting starts or re-starts. Table 15.4 shows the mode table for reset, load, count and hold actions.

Table 15.4 The mode table for the 74160 counter

Mode	Inputs						Outputs	
Pins	MR	CP	CEP	CET	PE	D	Q	Tc
Reset	0	X	X	X	X	X	0	0
Parallel	1	↑	X	X	0	0	0	0
load	1	↑	X	X	0	1	1	*
Count	1	↑	1	1	1	X	X	*
Hold	1	X	0	X	1	X	Qn	*
	1	X	X	0	1	X	Qn	0

Parallel loading is enabled when the PE pin voltage is taken to 0, and in this state each Q output will take the state of its corresponding D input. Figure 15.9 shows a five-decade counter (to a count of 99999) which makes use of these units, connecting them together so that synchronous counting is preserved. Note that synchronous decade counters can be connected to each other in an asynchronous way, with the output of a counter driving the clock of the next, but the circuit shown here preserves fully synchronous action, using one counter to gate the following counters.

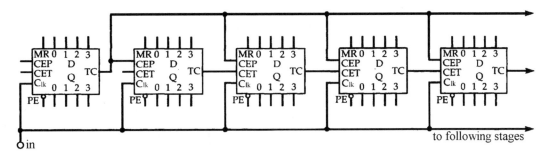

Figure 15.9 A five-decade counter using 74160 chips

Suppose that we want to count in scales other than 8-4-2-1 binary or BCD? The gating methods that are used for synchronous counters allow for any counting system to be used, and design is carried out making use of state tables to show what J and K entries are required, and diagrams called *Karnaugh maps* to show what gating can achieve this. The topic of Karnaugh maps is one of more importance to designers than to service engineers, and is not included here.

The important point is that whereas a binary count can be achieved to any count number by repeating a basic design, this is not necessarily true of other types, and unless a designer can work with very large Karnaugh maps, a practical limit of four stages is reasonable. As it happens, most unorthodox counters are four-stage counters in any case.

Activity 15.4

Assemble and test modulo-n (where n is any whole number) counters, both synchronous and asychronous. Use ICs such as 74LS74, 74LS112, 74LS193 and 74LS390.

Shift registers

A register is formed by interconnecting a number of bistables in a similar way to a counter. Each bistable forms a memory cell for one bit of information. The data may be input in serial or parallel form and read out in a similar manner, depending on the method of interconnection. There are thus four basic forms of shift register:

Serial in-serial out (SISO).

Serial in-parallel out (SIPO).

Parallel in-parallel out (PIPO).

Parallel in-serial out (PISO).

Some registers are of one form only whilst others are more universal, with the mode of data transfers controlled by the logic values of separate control lines. The general principle is explained using Figure 15.10, where a simple basic register is formed from D-type flip-flops (R-S flip-flops may also be used). Remember that the logic value on the D input is transferred to the Q output on the next clock pulse. In this example, each data value input thus ripples through the register in a manner similar to the counter.

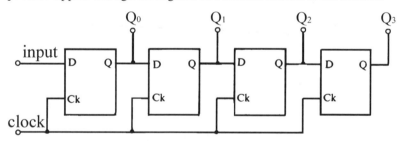

Figure 15.10 A simple four-bit shift register

Question 15.2

$Q_0 = 1$ $Q_1 = 1$
$Q_2 = 0$ $Q_3 = 1$
$D = 0$

The current state of the logic values for the shift register of Figure 15.10 is as shown, left.

What is the state of the Q outputs (in descending order of significance) after the next clock pulse?

The functions performed by shift registers include:

- Multiplication and division (SISO).
- Delay line (SISO).
- Temporary data store or buffer (PIPO).
- Data conversion; serial to parallel and vice versa (SIPO and PISO).
- Serial input-parallel output shift register

Figure 15.11 shows a hard wired version of a 4-bit shift register based on D-type flip-flops with AND-OR control logic. This provides for parallel data loading with serial format output and, in addition, parallel access to the same data stream.

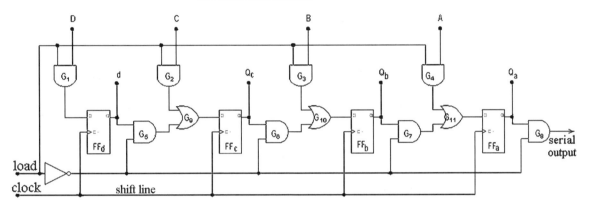

Figure 15.11 Parallel-in, serial-out (PISO) shift register with parallel access

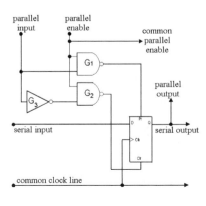

With the *load* line set to logic 1, gates G_1, G_2, G_3 and G_4 are enabled and any input at logic 1 is loaded into the D inputs. Similarly, any inputs set to logic 0 will simply provide a zero input. At the same time, gates G_5, G_6, G_7 and G_8 will be disabled to prevent any data movements. When the load line is set to logic 0, the serial gates are enabled and the load gates disabled. The data bits can now be transferred to the Q outputs on the negative going edge of each succeeding clock pulse. After four clock pulses the shift register is again ready to be loaded with the next nibble.

The diagram, left, shows an alternative approach to achieve parallel loading of a D-type shift register. This uses the input logic level to force the Q output to take up the same state via the preset and clear inputs. When the parallel load line and the data input is high, Pr is driven low and Clr high through the steering logic, to force the Q output to the logic 1 state. If the data input is at logic 0, the states of Pr and Clr are reversed and the Q output is forced to logic 0. Once loaded, the parallel load line is disabled and the register data is right shifted on the negative going edge of each clock pulse.

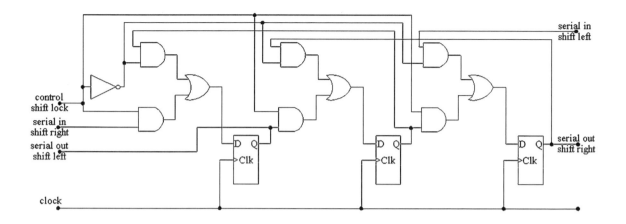

control
shift lock
serial in
shift right
serial out
shift left

serial in
shift left

serial out
shift right

clock

Figure 15.12   3 stages of an 8-stage shift left/shift right register

Figure 15.12 shows a simplified diagram of part of an 8-bit dual purpose shift register. In the practical device, the central flip-flop and its control logic is duplicated as often as needed. With the shift control line set to logic 1, all the lower AND gates are enabled so that data can be right shifted at each clock pulse. Conversely, with the control bit set to 0, the upper AND gates are enabled so that data can be left shifted.

Activity 15.5

Using ICs such as the 74LS194, assemble and test shift registers of each basic type.

Ring counters

Since a register is a set of flip-flops, usually connected so that one clock pulse can be applied to each of the flip-flops in the register, each register is potentially a synchronous counter. Counters constructed from registers, however, are not the normal type of binary counter but more usually the one-digit-per-device type, as illustrated in Figure 15.13.

This shows a set of flip-flops connected in the usual Q to J, Q# to K, mode, and with a reset arrangement which resets all but one of the flip-flops, placing a logic 1 into the other remaining flip-flop. After resetting, each clock pulse will then circulate that 1 by one place to the right and so back to the first flip-flop again. This is *not* a binary count, and the number scale that is used depends directly on the number of flip-flops that are connected in this way, a *straight-ring* connection.

A denary counter can be made using ten flip-flops in each decade, with the zero output used to pass a pulse to the next (higher order) set. This allows pulses to be counted with a direct output to simple display devices, like lamps set behind numbers on transparent disks. Counters like this are more useful for non-denary systems, however, since there are ample denary counters in IC form available. Ring counters come into their own where counts of 3, 4, 7, 9, 12 and others are required along with simple display devices. The straight-ring counter requires as many flip-flops as there are digits in the count, ten for denary, five for a scale of 5 and so on.

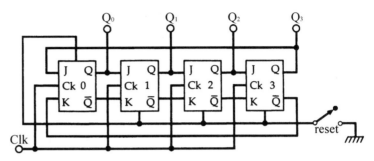

Figure 15.13 A straight-ring counter circuit using a SIPO register. The register must be set up with one 1 bit, since it cannot count if all outputs are zero

Twisted-ring counter

The twisted-ring or Johnson counter is a design that has some peculiar advantages for specialized purposes. Like a Gray code counter, it makes use of a single digit change at each count change, but unlike Gray code it does not make full use of the number of states that is possible.

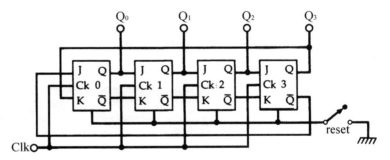

Figure 15.14 A twisted-ring or Johnson counter using four flip-flops

The count of a Johnson counter is $2 \times n$, where n is the number of stages, rather than the 2^n that can be achieved with true binary (8-4-2-1 or Gray) counters. A four-stage Johnson counter will therefore give a scale of nine rather than the scale of 16 that a true binary counter can provide. The Johnson numbers always consist of a block of 1s and a block of 0s with no alternation of digits. A Johnson count for a four-stage counter is shown in Table 15.5 for the shift-register connections illustrated in Figure 15.14.

The diagram of Figure 15.14 shows the reason for the name *twisted-ring counter* because it contains one connection of Q to K and Q# to J among a set of Q to J, Q# to K connections. The ability to create any count sequence is the most valuable feature of synchronous counters using JK flip-flops, and the methods that have been illustrated in this chapter can be applied to any count type.

Table 15.5   Count table for four-stage twisted-ring counter

Pulse No.	Q_0	Q_1	Q_2	Q_3	$Q\#_3$
0	0	0	0	0	1
1	1	0	0	0	1
2	1	1	0	0	1
3	1	1	1	0	1
4	1	1	1	1	0
5	0	1	1	1	0
6	0	0	1	1	0
7	0	0	0	1	0
8	0	0	0	0	1

At the 9th pulse, conditions return to those of pulse 1.

Question 15.3

The Johnson counter can produce a higher count from a given number of stages, but a binary or a straight-ring counter is usually preferred. Give one reason.

As it happens, most established counting scales present no problems, but you might want to use some very odd scales indeed. Digital to analogue conversion from BCD, for example, might require a counter that gave, for counts of 0 to 9, a number of 1s equal to the count number (five 1s for 5, eight 1s for 8 and so on). This is a modification of the Johnson idea which requires one flip-flop for each 1. This would normally be considered rather wasteful of devices, but if it's the easier way out of a problem then it will be worth doing. Though the 74 series of ICs contains a large variety of counters, there is always likely to be an application for unusual designs for

Applications of counters

particular needs, and for such actions, the construction of a synchronous counter using JK flip-flops is often the simplest solution that involves the least amount of hardware.

Apart from direct pulse counting, counter devices can perform many functions in industrial process control. Objects on a conveyor belt can be counted using a photo-electric cell and a light source. The reset pulse can be used to initiate an action to close a bottle after it has been loaded with the preset number of pills.

Downward count. A downward count to zero is also useful. This is used on some automatic coil-winding machines. The required number of turns to be wound on to a bobbin is first preset, and the machine set in motion. When the count reaches zero the machine automatically comes to a halt.

Frequency. A counter can be used to measure frequency of both unipolar (d.c.) and bipolar (a.c.) waveforms by counting the number of cycles over a precisely controlled period of time. It can also be used to measure time if driven by a precisely known crystal frequency.

Timekeeping. This is the basis of the digital watch or clock. In radar or sonar systems a transmitted pulse is reflected from an object and some of the energy returned to the source. The time delay can be measured and since the velocity of radio or sound waves is known, the time can be converted into distance.

Velocity. Two photo-cell/light source pairs can be set up a known fixed distance apart. The time taken for an object to pass between them can be measured and the distance/time computed to give the velocity.

Activity 15.6

Use a counter circuit and a plain toggle circuit to show that switch bounce causes a miscount. Try a variety of different switches.

Figure 15.15 shows the basic principles of a counting device that has been adapted to measure frequency. A crystal controlled oscillator frequency is divided down to produce an accurate time gating signal. In practice, the frequency is chosen to be high enough to provide the maximum count or minimum time ranges to meet the user demand.

In the case shown, the basic frequency is 1 MHz which when divided by 10^6 provides a time gate of 1 second. The input signal is converted into a square wave by suitable circuit and then input to the AND gate which is enabled by the timing gate signal. The displayed count then represents the unknown frequency in cycles per second or Hz.

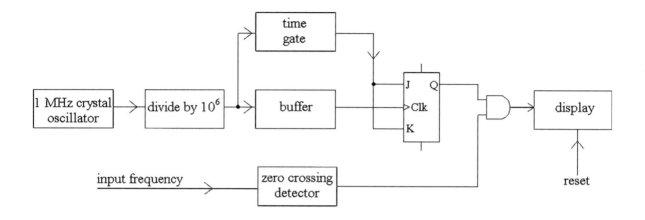

Figure 15.15   Frequency measuring counter

Figure 15.16 shows how the same basic components can be organized to produce a timer circuit. The crystal controlled oscillator output is again divided down to produce a suitable enabling signal to the AND gate which is further controlled via a Set/Reset (S/R) bistable device. When set and reset by start and stop signals, the display output represents the time delay t, between the two pulses. For example, if the decade divider output of 100 kHz is selected and a count of 87 is measured, then the time $t = 10^{-5} \times 87 = 0.87$ms. A positive-going start pulse is applied to the set input, and a positive-going stop pulse to the reset input.

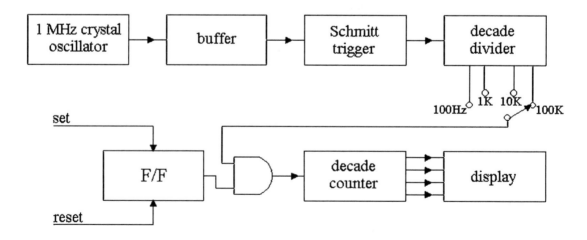

Figure 15.16   Time interval measurement by pulse counter

Answers to questions

15.1 The forbidden state is now R = 1, S = 1, and the latch state is R = 0, S = 0.

15.2 0101.

15.3 The output of a straight-ring counter can be displayed without the need for decoding.

16 Test equipment and fault finding

Servicing instruments and methods

In some respects, servicing logic circuitry can be simpler than working on analogue circuits of comparable size. All digital signals are voltages at one of two levels, and there is no problem of identifying minor changes of waveshape which can so often cause trouble in linear circuits. In addition, the specifications that have to be met by a microprocessor circuit can be expressed in less ambiguous terms than those that have to be used for analogue circuits. You don't, for example, have to worry about harmonic distortion or intermodulation, and parasitic oscillation is rare, though not impossible.

That said, logic and microprocessor circuits bring their own particular headaches, the worst of which are the relative timing of voltage changes and the difficulty of displaying signals. The actions of any microprocessor circuit depend on strict timing of many pulses being maintained, and conventional equipment which serves well for the analysis of analogue circuits is of little use in working with microprocessor circuits. The problem is compounded by the fast clock rates that have to be used for many types of microprocessors.

You may, for example, be looking for a coincidence of two pulses with a 33 MHz clock pulse, with the problem that the coinciding pulses happen only when a particular action is taking place. This action may be completely masked by many others on the same lines, and it is in this respect that the conventional oscilloscope is least useful. Oscilloscopes as used in analogue circuits are intended to display repetitive waveforms, and are not particularly useful for displaying a waveform which once in 300 cycles shows a different pattern. A fast conventional oscilloscope is useful for checking clock pulse rise and fall times, and for a few other measurements, but for anything that involves bus actions a good storage oscilloscope is needed.

Software and hardware, such as that provided by Pico Technology, can be used along with a fast modern PC computer, to provide facilities for logic circuit diagnosis, see Chapter 7.

In addition, some more specialized equipment will be necessary if anything other than fairly simple work is to be contemplated. Most of this work is likely to be on machine control circuits and the larger types of computers. Small computers do not offer sufficient profit margin in repair work to justify much diagnostic equipment. After all, there's not much point in carrying out a £500 repair on a machine which is being discounted in the shops to £450! We shall start this chapter looking at some typical fault conditions, and then at specialized instruments for digital circuit work.

Activity 16.1

Use suitable test equipment to locate faults in remote controls, A–D and D–A conversion circuits, and seven-segment displays.

Question 16.1

Why is a simple oscilloscope inadequate for trouble-shooting logic systems?

Common problems

Some of the problems that occur in digital equipment are due to poor servicing or poor manufacturing techniques, others may be brought about by failure of associated equipment (such as a regulated power supply). In this section we shall concentrate on common causes of failure of ICs. These are:

- incorrect insertion into sockets,
- pins shorted during measurements,
- poor soldering techniques,
- insertion/removal with power applied,
- incorrect voltage levels,
- input pins left disconnected,
- electrostatic discharge.

Incorrect insertion

Incorrect insertion into a socket is most likely for the larger DIL ICs, and the most common error is that pin 1 of the IC has been put into the pin 3 position of the socket or pin 3 of the IC has been put into the pin 1 position of the socket. Since the V+ and Earth pins are usually located at opposite corners of a chip, incorrect insertion will almost certainly make one of these connections o/c. The older STTL chips (and many of the LS types) will survive having power switched on in these circumstances, but MOS chips are more vulnerable. Always be prepared to test and, if necessary, replace a chip that has been incorrectly inserted.

Shorting pins

Making measurements of voltages on a working chip is reasonably simple for chips of the 74 family type, provided the correct tools are used. By far the best method is to slip an adapter over the chip. These are made for the most popular pin layouts, and they provide a set of contacts on the top cover that correspond to the IC pins and are electrically connected through clips. These contacts are well away from the PCB surface, and you can make clip connections to them without any risk of short circuits.

Without such adapters, great care has to be taken over pin voltage measurements, particularly the voltage difference between pins. Absolute voltage readings can be taken by clipping one lead of the meter to an earth point, and using a probe held to the IC pin. The probe should be insulated almost to the contact point to prevent shorting between pins.

By far the worst measuring problems relate to closely-spaced pins, particularly the square layouts used on microprocessors and ASICs. Adapters are available from the manufacturers for some types, but the only alternative is to use a fine probe with insulation almost to the tip.

Poor soldering

Poor soldering is a problem that is not quite so common now that manufacturing techniques have improved, but it is still a source of faults. The nature of modern boards with very fine tracks, small solder-pads and surface mounted (SM) components requires excellent soldering systems, and poses a problem for fault-finding because a badly soldered joint is very difficult to detect by eye.

A good magnifying glass and a bright light can help, but if you are convinced that the problem is in an area that appears to be all right, the usual treatment is to apply a soldering iron to see if this clears up the problem. The iron should be fitted with a very small bit, and it often helps to give the suspect area a thin coating of an approved flux paint. Needless to say, the equipment must be switched off and time allowed for capacitors to discharge before any such re-soldering is attempted.

Another aspect of this problem is a poorly soldered repair. Even with care, some repairs on modern PCBs are very difficult, particularly with SM components, and soldering must be suspected if a circuit fails to operate correctly after a repair (involving soldering) has been carried out.

Power-on working

We tend to take it for granted that power will be off when any repair is being done, but there is a temptation to remove and insert chips with power still on. This is something that must be resisted, even if switching off and on is not a straightforward action (as for many computer boards). Not only must the power be off when any chip is removed or replaced, it must have been off for long enough for capacitors to have discharged.

- Remember that some PCBs may contain back-up batteries, and these may have to be temporarily removed for some repairs. Since the backup batteries are there to keep memory working, this will require a considerable amount of work after restoring the battery, and you should consult the manufacturer's recommendation on this subject. If you need to replace a battery on a PC computer motherboard, you may be able to connect an external battery to maintain the memory while the internal cell (nowadays often a lithium-ion type) is replaced. If not, motherboards usually provide for a set of defaults to come into use after the backup battery power has been interrupted.

Incorrect voltage levels

Incorrect voltage levels can arise in several ways. One is the incorrect connection of cables that carry power from a PSU to a board, another is the failure of regulation of a power supply. Incorrect cable connection is unlikely on units such as modern PCs (using the ATX motherboard system), but if there is any doubt, the alignment of the connectors should be marked before they are separated.

Failure of regulation is a problem that is less easily avoided. Some circuit boards are surprisingly immune to regulation failure, particularly when no signals are applied, but for complex circuits, the PSU should incorporate over-voltage protection systems.

> Any work that is done on the PSU should preferably be done with the unit separated, and then with a dummy load connected for testing. In this way, any faults that cause incorrect voltage outputs will not damage ICs on the main boards.

Disconnected inputs

Input pin disconnections can arise from the incorrect insertion of ICs into sockets or from incorrect soldering of ICs to a board. They cause little harm on STTL circuits, where a disconnected input will *float high*, but can cause damage on MOS ICs when power is applied. Modern circuit boards will be designed so that a path to earth is available on any input, but only if the IC pins are connected to their tracks.

Electrostatic discharge

Electrostatic discharge should never be a problem for any IC that is correctly connected into circuit on a board, because the board will provide discharge paths. Damage is more likely when ICs are handled prior to insertion in a board, because the only protection available is by way of diodes incorporated into the chips, and these offer only limited protection.

The comparatively high humidity of the climate in the UK allows us to get away with working practices that would cause electrostatic damage elsewhere, and it is quite common even in manufacturing to see MOS devices being handled without elaborate electrostatic precautions. This is not, however, good practice, and the basic precautions of working in a well-earthed environment and keeping one hand earthed should be observed.

> The worst hazards are the ever-present plastic bags and other plastic packaging materials, all of which are electrostatic hazards. Computer components are always packed in conductive plastic (brown or black), and a collection of these materials can be useful to provide a clean conducting surface for placing sensitive ICs.

Question 16.2

In a digital voltmeter, why is a slow clock used to reset the action at intervals?

Avoiding electrostatic damage (ESD)

Electrostatic damage, see also Chapter 9, is most likely to affect MOSFETs, some of which can be damaged by a voltage as low as 35 V. Damage to a simple transistor is likely to cause failure, but damage to elaborate MOS ICs may be less obvious, causing reduced performance and/or shortened life. Table 16.1 indicates the level of electrostatic voltages that can be generated under typical conditions.

The figures for high relative humidity apply fairly generally in the UK where the workshop is naturally ventilated, but the lower figures are more likely where air-conditioning is used.

Note that even bipolar devices and diodes can be damaged by less than 3 kV levels, though the amount of current that can pass is often too low to cause lasting damage.

Damage to MOS ICs is possible whenever an electrostatic voltage can be applied to a pin or set of pins and can cause current to flow to grounded pins. This is most likely if an IC is incorrectly inserted or soldered so that some pins are earthed and one or more unconnected.

Table 16.1   Electrostatic voltages in the workshop

Action	Relative humidity (RH)	
	Low (10–20%)	High (65–90%)
Walking over carpet (artificial fibres)	35 kV	1.5 kV
Walking across a vinyl floor	12 kV	250 V
Working at an unprotected bench	6 kV	100 V
Shuffling paper	7 kV	600 V
Picking up a polythene bag	20 kV	1.2 kV

Device sensitivity

The devices that are most sensitive to electrostatic damage are MOS and CMOS chips that have no internal protection circuits, and discrete MOSFETs that also have no discharge paths built in. The utmost caution should be exercised in handling such devices, observing all the precautions recommended by the manufacturers.

Less sensitive devices include 74LS chips, analogue ICs, r.f. transistors for 500 MHz or more, semiconductors that incorporate silicon dioxide insulation, and MOS/CMOS devices with diode protection for inputs, along with JFETs and precision thin-film resistors. To put all this into perspective, I have never had a chip damaged by ESD in 45 years of *handling* components. Damage to items such as LNB circuits is more likely to be caused by electrical storms.

The least sensitive devices include microcircuits, small-signal transistors (less than 10 W output) and thick film resistors.

ESD protection

Some of the protective measures that are normally used to prevent ESD damage are noted here.

- Before starting any servicing work, check technical manuals and any leaflets for warnings and instructions regarding electrostatic damage precautions. Ensure that the working environment is as free as possible from electrostatic hazards, and try to ensure that nothing you do will generate high electrostatic voltages (like shuffling your feet on a carpet or on vinyl tiles).

- Before servicing equipment, you should earth yourself. The simplest way is to touch an earthed metal object, but if you have acquired a substantial amount of charge this can be painful. A better method is to provide in the workshop an earthed rail or wire to which 1M resistors are connected at intervals, with a small metal ball soldered to the free end of each resistor. Touching one of these metal balls will earth you painlessly without sparks.

- In low humidity conditions it is much more satisfactory to wear an earthing strap, a metal strap that is permanently connected (usually through a 1M resistor) to earth.

Safety regulations demand that you should not use any mains-operated equipment while you are wearing an earthing strap.

- The most risky procedures involve the unpacking of new devices. Handling must be kept to a minimum, with attention to earthing at all times, and you should try to ensure that the workshop environment is maintained so that materials that cause high electrostatic voltages are kept to a minimum. Do not unpack sensitive devices until you are ready to use them.

- The conductive packing plastic that is used for sensitive devices can be used to hold the chips while you insert them, but if this is not possible, hold ICs by the body corners only, avoiding any contact with the pins. Avoid in particular any contact between IC pins and any clothing or (non-conducting) plastic. If you are not wearing a wrist strap, try to keep contact with an earthed surface while you are handling a sensitive IC. Avoid touching synthetic materials that acquire electrostatic charges. Use soldering irons that are earthed, and avoid the use of solder suckers that have plastic (PTFE) tips, unless the tips are guaranteed to be antistatic.

Activity 16.2

Use suitable test equipment to locate faults in synchronous and asynchronous counters, shift registers and bistable circuits.

Test equipment

Test equipment for digital circuitry must, of necessity, include some items that are not normally used on analogue circuits, but this does not mean that familiar instruments such as the multimeter are used to any less an extent. A good analogue or digital voltmeter is a very valuable backbone of all modern servicing, but you should know its limitations.

When the analogue meter is used for voltage readings you need to know the resistance of the meter for each voltage range. The meter resistance can be found from the *ohm-per-volt* figure, or *figure-of-merit*, which is printed either on the meter itself or in its instruction booklet. To find the resistance of the meter on a given voltage range, multiply the figure-of-merit by the voltage of the required range.

Example: What is the resistance of a 20 KΩ/V voltmeter on its 10 V range?

Solution: Meter resistance is 20K × 10 = 200K on the 10 V range.

If you need to take a reading across a high resistance, a high-resistance meter range must be used. You may be able to use a higher meter range than the one which seems to be called for. For example, if a voltage of around 9 V is to be measured and the 10 V range of the meter has too low a resistance, the 50 V or even the 100 V range can be used to reduce the distorting effect of the meter on the circuit. Note, however, that this would not be possible if the reading on the 100 V range would thereby be made too low to read. A voltage of around 1.5 V or less would be unreadable on the higher-range scales.

Digital multimeters generally have much less effect on the circuit being measured. Most of them have a constant input resistance of about 10M or more, and few circuits will be greatly affected by having such a meter connected into them. The operating principle is that the input voltage is applied to a high-resistance potential divider which feeds a comparator (see Figure 16.1).

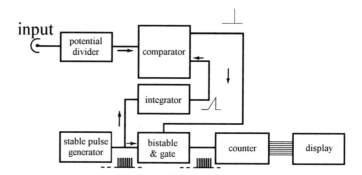

Figure 16.1 Block diagram of a digital meter omitting details of hold-and-measure arrangements

Another part of the circuit generates a sawtooth wave which is applied to the other input of the comparator. At the time that the sawtooth starts, a counter also starts and is stopped when the two voltages at the inputs of the comparator become equal. At that moment the count is displayed. The circuit is arranged so that each digit of the count corresponds to a unit of voltage, say one millivolt, so that the display reading is of voltage. The range switch selects the part of the potential divider to be used, and the position of the decimal point on the display.

For example, suppose that the pulse generator or clock of the digital meter runs at 1 kHz, so that 1000 pulses per second are generated. If the integrator is arranged so that each pulse produces a voltage rise of 1 mV, then the voltage will rise to 1 V in the time of 1 second. For an input of 0.5 V, 500 pulses will be needed, and the time needed will be 0.5 s. The display will then indicate a reading of 0.500 V. Note, however, that the instrument has taken 0.5 seconds to reach this reading.

Note also the following points:

- In general, digital meters are not capable of following rapid changes in voltage.

- Although a digital meter may indicate a voltage reading to several places of decimals, this is not necessarily more precise than the reading on an analogue meter.

- Meters with still higher resistances are also available up to several thousand megohms, if required.

Current tracer

Work on printed circuit boards makes it difficult to trace currents, since it would be unacceptable to split a track to check the current flowing through it. A current-tracer (also called current checker) is a small hand-held device that makes it possible to detect the presence and direction of current flow on a PCB track, and also to indicate, roughly, its size.

The principles of one type are illustrated in the block diagram, left. Two probes, set a fixed distance apart, are pressed on the track, and the tiny voltage difference that is produced by the current flow is amplified by balanced operational amplifiers so as to operate LED indicators. Typically, three LEDs are used to indicate currents of 10, 50 and 100 mA on 1 mm track width, or 20, 100 and 200 mA on 2 mm track width. A fourth LED is used to indicate reverse polarity, or to indicate open-circuit track.

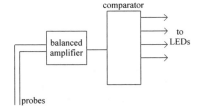

An alternative method of tracing current uses a Hall-effect probe which does not rely on sensing voltage, using instead the magnetic field around the track. This type of tracer, such as the *ToneOhm*, manufactured by Polar Instruments, will detect partial and complete short circuits and, with practice, can be used to find the position of an open-circuit.

A current tracer, like a voltmeter, is a useful first-test instrument, as it will assist in finding any problems that cause a drastic change in the current flowing along a track. This can help to pin-point damaged tracks (a hazard particularly in flexible circuit strips) or failure of components that draw large currents, such as printer-head driver transistors. One point to note, however, is that modern PCBs with narrow tracks make the use of any type of current tracer very much more difficult.

Logic probe

A logic probe is a device which uses a small conducting probe to investigate the logic state of a single line. The state of the line is indicated by LEDs, which will indicate high, low or pulsing signals on the line. The probe is of very high impedance, so that the loading on the line is negligible. Figure 16.2 shows a block diagram for a typical instrument.

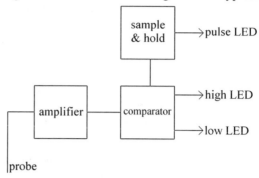

Figure 16.2   A simplified block diagram for a logic probe

The voltage on the track is amplified, and applied to a comparator which will detect high or low state. If the voltage is intermediate, as can be caused by the presence of pulses, the sample and hold circuit will rectify and store the pulse level, activating the third LED.

A typical logic probe can be switched to either TTL or CMOS voltage levels (+5 V supply for TTL and 3 V–18 V for CMOS) and uses coloured LEDs to identify the logic state of the track. Pulses as narrow as 30 ns can be detected, and as wide as 500 ms. Some types can also indicate the presence of ripple on the power supply, indicating faulty stabilization. Probes can be battery-powered, or can take their power from the circuit under test, using crocodile clips.

These probes are not costly, around £15 to £50, and are extremely useful for a wide range of work on faults of the simpler type. They will not, obviously, detect problems of mistiming, but such faults are rare if a circuit has been correctly designed in the first place. Most straightforward circuit problems, which are mainly chip faults or open or short circuits, can be discovered by the intelligent use of a logic probe, and since the probe is a pocket-sized instrument it is particularly useful for on-site servicing.

Obviously, the probe, like the voltmeter used in an analogue circuit, has to be used along with some knowledge of the circuit. You cannot expect to gain much from simply probing each line of an unknown circuit. For a circuit about which little is known, though, some probing on the pins of the microprocessor can be very revealing. Since there are a limited number of widely-used microprocessor types, it is possible to carry around a set of pinouts for all the microprocessors that will be encountered.

Starting with the most obvious point, the probe will reveal whether a clock pulse is present or not. Quite a surprising number of defective systems go down with this simple fault: it is even more common if the clock circuits are external. Other very obvious points to look for are a permanent activating voltage on a HALT line, or a permanent interrupt voltage, caused by short circuits. For an intermittently functioning or partly functioning circuit, failure to find pulsing voltages on the higher address lines or on data lines may point to microprocessor or circuit-board faults.

For computers, the description of the fault condition along with knowledge of the service history may be enough to lead to a test of the line that is at fault. The considerable advantage of using logic probes is that they do not interfere with the circuit, are very unlikely to cause problems by their use, and are simple to use. Some 90% of microprocessor system faults are detectable by the use of logic probes, and they should always be the first hardware diagnostic tool that is brought into action against a troublesome circuit.

Logic clip

The logic clip is an extension of the logic probe to cover more than one line. As the name suggests, these devices clip over a logic IC, and are available for 14-pin or 16-pin DIL packages. As the block diagram of Figure 16.3 shows, each pin of the IC is connected to a buffer in the logic clip, and this buffer drives an LED indicator. Logic clips are usually available in separate TTL or CMOS versions, though the TTL version is now more common because of the widespread use of 74HC devices.

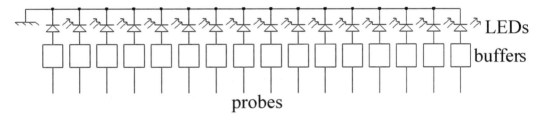

Figure 16.3   Block diagram for a logic clip

The logic clip is particularly useful when the system clock rate can be slowed down, or a logic pulser is being used to supply an input. Because all the states on a single IC chip can be monitored together, any fault in a gate within the chip is fairly easy to find, much easier than the use of a logic probe on all signal pins in turn.

Logic pulsers

Logic pulsers (or digital pulsers) are the companion device to the logic probe. Since the whole of a microprocessor system is software operated, some lines may never be active unless a suitable section of program happens to be running. In machine control circuits particularly, this piece of program may not run during any test, and some way will have to be found to test the lines for correct action.

A logic pulser, as the name indicates, will pulse a line briefly, almost irrespective of the loading effect of the chips attached to the line. The injected pulse can be detected by the logic probe. This method is particularly useful in tracing the path of a pulse through several gate and flip-flop stages. A typical block diagram is illustrated in Figure 16.4.

The multivibrator generates a square wave whose frequency can be varied in the range 0.5 Hz to 400 Hz, and this wave is used to trigger a pulse generator which provides positive and negative pulses of about 10 µs width. The output stage will float to the voltage of the line that the probe is touching, and can source or sink up to 100 mA. This is enough to override any other drivers on the line. A synchronizing pulse can be used to trigger an oscilloscope if needed.

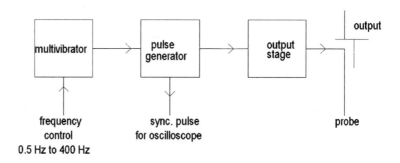

Figure 16.4   A typical block diagram for a logic pulser

The logic pulser is a more specialized device than the logic probe, and it has to be used with more care. It can, however, be very useful, particularly where a diagnostic program is not available, or for testing actions that cannot readily be simulated.

Logic analysers

The logic analyser is an instrument which is designed for much more detailed and searching tests on digital circuits generally and on microprocessor circuits in particular. As we have noted, the conventional oscilloscope is of limited use in microprocessor circuits because of the constantly changing signals on the buses as the microprocessor steps through its program. Storage oscilloscopes allow relative timing of transitions to be examined for a limited number of channels, but suitable triggering is seldom available. Logic probes and monitors are useful for checking logic conditions, but are not helpful if the fault is one that concerns the timing of signals on different lines.

The logic analyser is intended to overcome these problems by allowing a time sample of voltages on many lines to be obtained, stored, and then examined at leisure. Most logic analysers permit two types of display. One is the 'timing diagram' display, also called 'timing domain analysis' in which the various logic levels for each line are displayed in sequence, running from left to right on the output screen of the analyser. A more graphical form of this display can be obtained by connecting a conventional oscilloscope, in which case, the pattern will resemble that which would be obtained from a 16-channel storage oscilloscope. The synchronization may be from the clock of the microprocessor system, or at independent (and higher) clock rates which are more suited to displaying how signal levels change with time.

The other form of display is word display or 'data domain analysis'. This uses a reading of all the sampled signals at each clock edge, and displays the results as a 'word' for each clock pulse rather than as a waveform. If the display is in binary, then the word will show directly the 0 and 1 levels on the various lines. For many purposes, display of the status word in other forms, such as hex, octal, denary or ASCII, may be appropriate. This display, which gives rise to a list of words as the system operates, is often better suited for work on a system that uses buses, such as any microprocessor system.

The triggering of either type of display may be at a single voltage transition, like the triggering of an oscilloscope, or it may be gated by some preset group of signals, such as an address (a trigger word or event). This allows for detecting problems that arise when one particular address is used, or one particular instruction is executed. One common method is to trigger a display by using a combination of inputs. One of these would be a trigger word which can be set and stored, the other inputs would be trigger signals (qualifiers) which can be taken from the clock (a clock qualifier) or from other inputs (trigger qualifier). You can also use a word search (trace word or event action) through the memory of the analyser to find if a specified word has been stored in the course of an analysis. Figure 16.5 shows a simplified block diagram of a typical analyser.

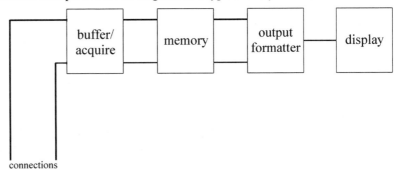

Figure 16.5   Simplified block diagram for a logic analyser

The data acquisition portion of an analyser is illustrated in the block diagram. The logic levels at the 16 inputs are sampled, using the internal or an external clock pulse to synchronize the sampling. These levels are stored, and in the usual operating mode, the triggering will cause the stored data to be centred around the triggering time.

For example, there may be 2K of data both before and after the trigger event. Triggering is an ANDed action so that you can set for some combination of signals that is unique, such as when the microprocessor writes a specified word to a chip during an interrupt.

Note that a display such as can be achieved using a logic analyser can also be obtained from a computer simulator. Simulator software allows the user to notify the chips and connections that are used in a digital circuit, and the program will then provide simulated waveforms. The value of this system is that any unwanted pulses (called 'glitches') can be detected, even to the extent of a 1 ns pulse that in real life would occur once in 14 days, in the proposed circuit before it has been constructed, and the simulator can also be used to find such weaknesses in an existing circuit. One well-known simulator is PULSAR, from Number One Systems Ltd. of St. Ives; this has the advantage that it can be integrated with circuit diagram and PCB layout software.

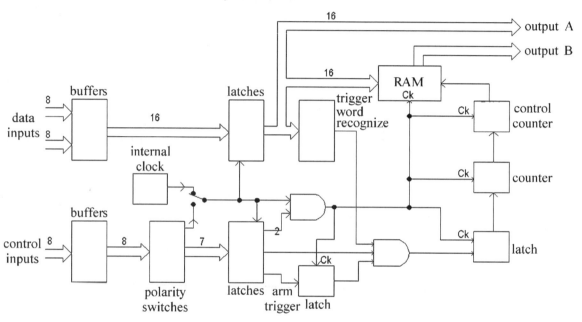

Figure 16.6   A simplified block diagram for the data acquisition portion of the Thurlby LA-160 analyser, by kind permission of Thurlby Electronics

Question 16.3

Describe how you would use (a) a logic probe, (b) a logic pulser.

Signature analyser

A *signature* in this sense means a hexadecimal word, and the signature analyser is a method of finding an error in any logic circuit, including the memory chips for ROM, EPROM or CMOS RAM. The action is that a hexadecimal word is produced from each output when a set of inputs is applied to all the possible inputs of the circuit.

The basic diagram of the essential portion of a signature analyser is illustrated in Figure 16.7. This consists of a serial register with feedback taken from four of its parallel outputs to an XOR input. The feedback is applied so that any typical set of inputs will produce a set of bits in the register which is almost certainly unique – the principle, based on information theory, is that a register is working most efficiently when all states are equally likely (an n-bit shift register produces $2^n - 1$ different output patterns).

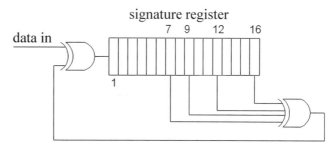

Figure 16.7   The register with feedback that is the basis of a signature analyser

Without feedback, a 16-bit register would produce a hex word that would simply reflect the last 16 bits fed into it. For example, if you fed in 16 1s, the register word would be FFFFH, and this would not alter if you continued to feed in 1s. By contrast, with the feedback connections illustrated, the action of feeding in 1s will produce a different register word for each 1 bit fed in, and there will be no repetition until 65536 bits have been fed in. This is illustrated in Table 16.2 (following) which shows 45 words that are generated when a stream of 1s is fed in – the first two words are omitted because they are simply 0001H and 0003H. The effect of the feedback starts to become evident after seven 1s have been fed in.

This register arrangement is also used to produce 'random' numbers, and when larger registers are used the chance of recurring numbers becomes less. In the 16-bit example, the words that are generated from 65 536 or fewer entries are therefore unique, and this allows us to trigger the start and stop of a signature analyser from the highest order of address line (A15) of an 8-bit microprocessor system, or from the A15 line of a system that uses more than 16 address lines.

Figure 16.8 illustrates the block diagram of a signature analyser with its inputs of clock, start, stop and data. The data input to the analyser will be from any line that produces signals, so that address or data lines can be used. Contact can be made by a probe or a clip. The register is also gated so that it operates for a fixed 'window', a defined number of clock cycles. In use, the system under test is 'exercised' meaning that inputs are applied – preferably so that all possible signal values are used at the inputs. The register of the signature analyser will fill with bits in the gated time window, and when the gates have closed, the four hex numbers displayed by the register form a word that should be unique to that combination of inputs and the point that is being tested.

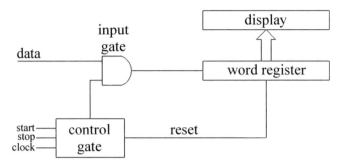

Figure 16.8   The block diagram of a signature analyser

Table 16.2   Register (hex) words generated by a stream of level 1 bits

No. of 1s	Word	No. of 1s	Word	No. of 1s	Word
3	0007	18	F9CC	33	72A2
4	000F	19	F399	34	E545
5	001F	20	E733	35	CA8A
6	003F	21	CF67	36	9515
7	007F	22	9CCE	37	2A2B
8	00FE	23	399C	38	5456
9	01FC	24	7339	39	A8AC
10	03F9	25	E672	40	5159
11	07F3	26	CCE5	41	A2B3
12	0FE7	27	99CA	42	4566
13	1FCE	28	3395	43	8ACD
14	3F9C	29	672A	44	159A
15	7F39	30	CE54	45	2C34
16	FE73	31	9CA8	46	5669
17	FCE6	32	3951	47	ACD2

The signature analyser must be used along with software that ensures that the inputs will be consistent, and a system that is known to be perfect is used to provide the signature words. If a system on test displays a different signature this will point to a fault, and the position of the fault can be found by repeating a signature check at different points along a signal path. One merit of signature analysis is that the same type of fault at a different point in a circuit will usually produce a different signature word.

> • Testing of combinational circuits is straightforward, but sequential circuits can be tested only if all feedback connections are broken.

Microprocessor circuits are usually signature tested by inserting an adapter between the microprocessor and its socket. The adapter is designed to allow pins to be open circuited or connected to either logic level so as to prevent interrupts or to isolate data lines from memory, and also to set a fixed data input such as the no-operation (NOP) code which will allow the clock to cycle the address lines continually. The probe can then be used at each point where a signature is known.

Figure 16.9 illustrates the modifications that have to be made to a Z80 microprocessor circuit in order to carry out signature analysis. The interrupt line is broken so that the interrupt pin can be taken high, and the data bus is also broken so that the NOP command is always on the microprocessor data pins. These changes can be incorporated into an adapter that fits between the microprocessor chip and its normal socket.

A more restricted meaning of signature is applied to the ROM of a computer. The signature can be found in a variety of ways, such as by adding up all the stored byte numbers and taking the remainder after dividing by some factor – whatever method is used will have been devised so that the resulting word is unique and will be found only if the ROM contents are uncorrupted. The word that is obtained in this way will also be stored in the ROM, so that if the two do not match there must be a ROM error. This system can be used also on RAM when the RAM is filled with specified bytes.

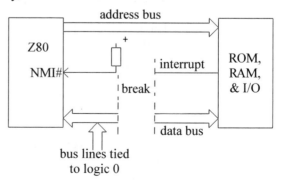

Figure 16.9   Signature analysis modifications to a Z80 circuit

Question 16.4

A ROM chip signature is quoted as 7F4BH. What does this mean, and how could you check it?

Activity 16.3

Use suitable test equipment to locate faults in systems that make use of combinational logic circuits.

Faultfinding techniques

Faultfinding for electronic/electrical equipment is a skill that is neither an art nor a science, but an engineering discipline in its own right. Effective faultfinding requires:

- A good general knowledge of electricity and electronics.
- Specialized knowledge of the faulty equipment.
- Suitable test equipment.
- Experience in using such test equipment.
- The ability to formulate a procedure for isolating a fault.
- The availability of service sheets and other guides.

A good general knowledge of electricity/electronics is an essential because not all equipment is well-documented, and in some cases only a circuit diagram (or even nothing at all) may be available as a guide. Failing a concise description of how the equipment works, you may have to work out for yourself the progress of a signal through the equipment. In addition, a wide general knowledge is needed if you are to make reasonable assumptions about how to substitute components. You are not likely to know why something doesn't work if you don't know what *does* make it work.

Specialized knowledge can greatly reduce the time spent in servicing, and if your servicing is confined to a few models of equipment you are likely to know common or recurring faults by their symptoms. All too often, however, service engineers are likely to have to struggle with unfamiliar equipment for a large portion of their time.

Suitable test equipment is essential. The days when a service engineer could function effectively with little more than a multimeter and a screwdriver are long gone, and though the multimeter is still an important tool (as also is the screwdriver) the service engineer needs at least one good general-purpose oscilloscope, along with signal generators, pulse generators, and more specialized equipment appropriate for the type of equipment he/she is working on.

Experience in using test equipment is also an essential. All test instruments have limitations, and you must know what these are and how you can avoid being hung up by these limitations. You must know which tests are appropriate for the faulty equipment, and what the result of such tests would be on equipment that was not faulty.

The ability to formulate a procedure for isolating a fault means that you need to know what to test. All electrical and electronic equipment consists of sections, and much modern electronics equipment uses a single IC per section. You should be able to pin down a fault to one section in a logical way, so that you do not waste time in performing tests on parts of the circuit which could not possibly cause the fault.

- An important point about all fault-finding is that you should make note of all the tests you have applied and the results of these tests. Without this, you are likely to forget what you have done, and end up taking measurements haphazardly, making diagnosis impossible.

Do not forget the simple things. We tend to make fun of the customer who has forgotten to check if the mains switch was on, but we can be guilty of similar neglect ourselves. In particular, always check plugs and connectors for any equipment that is not functioning. The next logical check is that power supplies are present within the equipment. Once these elementary checks have been made, we can start looking for more serious faults.

The classical method of isolating a fault has, in the past, been to check signal inputs and outputs for each stage, but this is no longer the only method that needs to be used, and in some cases, the use of feedback loops, limiters, and other interacting circuits makes it much more difficult to find where a fault lies. Once again, experience is a valuable guide.

The availability of service sheets and other guides is also important. Much commercial equipment consists of components which carry only factory codes, and whose actions you can only guess at in the absence of detailed information. In addition, good service sheets will often carry a list of known recurring faults, and will also give valuable hints on fault-finding methods.

End to end technique

Basic fault finding in both analogue and digital systems follows principles that are similar. A source is required to inject suitable signals into the input and the signal processing is then monitored as it passes through the system on a stage by stage basis. For analogue systems a suitable input source is a signal generator, while an oscilloscope can be used as a monitor. For digital systems this *end to end* technique can be carried out using a logic pulser to provide the inputs while the processing can be monitored with a logic probe. You can use this method in either direction, input-to-output or output-to-input.

Half-split method

Another technique that is often used with advantage to speed up diagnosis is known as the *half-split method*. Here the system is divided into two sections and the end to end technique used to find the faulty half. Once this is found, this part is again divided into two, and the check repeated. This process is then repeated continually until the faulty stage is identified. It is unusual to require more than three repetitions of this method.

And, finally…

Finding a fault is not, unfortunately, a certain step towards repair. Some equipment carries ICs which are no longer in production and for which no replacement is available. Many firms, particularly manufacturers of domestic electronic equipment, will provide spares and help for only a limited period, and some firms seem to deny all responsibility for what their equipment does after a few years. Given the comparatively long life of most electronic equipment, it would be unreasonable to expect spares to be available indefinitely, but it is not easy to tell a customer that the TV receiver he/she bought only six years earlier cannot now be serviced because it contains parts for which these is no current equivalent. Manufacturers might like to remember that customers tend to have long memories about such things – it certainly affects my judgement when I want to buy anything new.

Activity 16.4

Construct a 16-bit serial register (using two 8-bit chips) and add the feedback connections from outputs 7, 9, 12, and 16 into an XOR (or an adder with carry ignored) to as to produce the format of the signature analyser register. Find the signature words for different numbers of input pulses.

Answers to questions

16.1   It can display only one of two traces, operates at relatively low speeds, and has no storage action.

16.2   So that a changing voltage can be tracked.

16.3   The probe tip is placed on a bus line to indicate logic levels of 0, 1 or pulsing. The pulser is connected to a bus line to inject pulses so that their effect can be traced.

16.4   This is the result of a check-sum on all the ROM contents and can be checked by repeating the check-sum calculation using software.

Appendix 1

Certification and assessment structure

The NVQs/RVQs cover the general core studies for radio, audio, television, industrial electrical/electronic systems, refrigeration, domestic cookers and gas and other small appliances. The syllabuses are designed for the vocational education of new entrants to the servicing industry. They specifically meet the requirements for the under-pinning technical knowledge that is needed to obtain the appropriate award. The major difference between the two awards, NVQ and RVQ, which are considered to be of equal status, lies in the way in which candidates progress is accessed and accredited.

For the NVQ, the college or training organization, assesses both the under-pinning technical knowledge and practical competence. The overall progress may be moderated and accredited by a training authority such as EMTA. For the RVQ which is progressively replacing the C&G 2240 and EEB structures, the changes will only be marginal. The C&G will still examine, by multiple choice question papers, the under-pinning technical knowledge, whilst the assessment of the practical competence topics will be the joint responsibility of the training organization and the EEB.

At all times, the health and safety aspect within both the work place and the area in which the systems are employed is specifically emphasized.

The Level 2 qualifications include five mandatory core units plus one optional unit. These candidates are expected to be able to work to clearly defined procedures and with responsibility for identifying and implementing any decisions. However, they will need to refer to other competent servicing personnel for authorization and guidance. Fault finding and repair for this award is restricted to module or stage level.

The Level 3 qualification requires a thorough understanding of four mandatory core modules plus two optional units. The candidates must demonstrate an ability to work to procedures that are not clearly defined. They must take responsibility for planning and controlling the quality of work, and to instruct and give guidance to other less well qualified servicing staff. Level 3 candidates must be able to demonstrate their ability to fault find and carry out repairs down to component level.

In addition to all this, all candidates must be able to demonstrate an ability to carry out soldering and de-soldering exercises and test for the serviceability of components.

Appendix 2

General syllabus structure for NVQ Level 3

The registration and certification for this course structure is a continuing process, with candidates' progress being monitored and tested and logged at each stage by appropriately qualified personnel. The *Contents* listing at the beginning of this book, reflects fairly accurately, the main unit headings covered by the syllabus for the common core technology needed for the Level 3 award.

For all disciplines of this new course, all candidates must have a good understanding of the *Health & Safety* regulations and the associated legislation as it affects the servicing industry. This aspect which affects both the workplace and the areas in which the equipment is employed, means that many of the specific applications can only be certified within the normal working environment.

Technically, each candidate is expected to demonstrate the ability to carry out testing and diagnosis down to stage or module level, including the correct dismantling and re-assembly. This includes the ability to recognize that the test equipment being used is operating correctly and is calibrated. After repair, the candidate must be able to re-install any equipment back in its original place of use to the satisfaction of the owner or user.

Particularly in an industrial environment, candidates are expected to be able to follow the instructions of any planned maintenance scheme and carry out the necessary updating of its database.

During the fault-finding exercises, the candidate is expected to be able to write a clear and concise report about the progression from first impressions to the diagnosis of the faulty stage or component. These reports are usually marked some time after the exercise, so that it is important that the candidate can not only demonstrate an ability to locate faults, but to describe the actions taken in order to deduce the cause of the problem, all in a logical manner.

- This is an exercise that requires a lot of practice. If only a little is written about the fault finding process, then very few marks are likely to be awarded.

The important steps can be summarized by the mnemonic TOC; Test, Observe, Conclusion. Note the system reaction to the switch on process, record what is seen and deduce what the next step should be. From the initial tests certain conclusions can be drawn and these lead logically to the next stage to test. Finally, after due consideration, the faulty stage and then the faulty component can be deduced in a logical fashion.

Appendix 3

Multiple-choice examination questions: guidance for examination candidates

The certifying and examining bodies for electronics/electrical servicing rely considerably on the use of multiple-choice questions at all levels. The sheer size of a multiple-choice question paper often leads candidates into feeling that time is short, leading to a feeling of urgency which results in bad or even silly choices being made. The first point to understand, then is that the time allocation is in fact quite generous, and the size of the paper is due to the space that is needed for printing diagrams and for the choice of answers.

The golden rule for answering this type of questioning is to formulate your own answer to the question before looking at the choice of answers, and then selecting the answer that most closely fits with your own. In this way, when you know the answer as you read the question you will not be distracted by reading over the list of choices, all but one of which will be designed to distract you. On your first reading of a paper, answer all of the questions that you can using this method, and only then proceed to look at the others.

If you are unsure of an answer of your own, then you can usually reject all but two of the possible answers that have been provided. If you are genuinely uncertain about the correct choice of these two, then you can opt for one or other knowing that your chances of being correct are considerably higher, 50 per cent instead of 25 per cent than they would have been if you had simply guessed in the first place.

A methodical approach like this can considerably improve any candidate's score in a multiple-choice examination, particularly when there are usually two or more possible distracting solutions that are most likely to be wrong.

Appendix 4

Examples of multiple choice test questions.

Chapter 1

1. When either a series or parallel tuned circuit is operated at resonance, the current flowing into the circuit is:
   (a) in phase with the applied voltage,
   (b) leads the voltage by 45°,
   (c) leads the voltage by 90°,
   (d) lags the voltage by 90°.

2. A 10 mH inductor has a reactance of approximately 31.42K at 500 kHz. Estimate the reactance of a 20 mH inductor at 1 MHz.
   (a) 62.84K,
   (b) 15.71K,
   (c) 125.66K
   (d) 94.26K.

3. A 1 nF capacitor has a reactance of about 3183 Ω at 50 kHz. The reactance of a 5 nF capacitor at 20 kHz would be approximately:
   (a) 15915 Ω,
   (b) 636 Ω,
   (c) 1591 Ω,
   (d) 7957 Ω..

4. An inductor and a capacitor are connected together in two different ways and the impedance measured at the resonant frequency. For circuit A the impedance at resonance is 5 Ω, for circuit B the impedance is 25K.
   (a) Circuit A contains a 5 Ω resistor and circuit B contains a 25K resistor.
   (b) Circuit A is a parallel circuit and circuit B is a series circuit.
   (c) Circuit A has the inductor connected the other way round as compared to circuit B.
   (d) Circuit A is a series circuit and circuit B is a parallel circuit.

5. An a.c. circuit contains a resistor, a capacitor and an inductor. The a.c.voltages across each component are measured, but the sum of the

voltages is not equal to the voltage across the whole circuit. This is because:

(a) such voltages never add up correctly,

(b) the phase angles are not being taken into account,

(c) voltmeters do not read correctly on a.c.,

(d) only the voltage across the resistor should be counted.

6. A large voltage can be developed across an inductor:

(a) when a steady voltage (d.c.) is applied across it,

(b) when a steady current (d.c.) flows through it,

(c) when a current flowing through it changes amplitude,

(d) after a current flowing through it falls to zero.

Chapter 2

1. A perfect transformer is defined by:

(a) primary voltage = secondary voltage,

(b) primary current out = secondary current,

(c) primary power out = secondary power,

(d) primary turns = secondary turns.

2. For a transformer with a 5:1 step-down ratio, a load of 1K in the secondary is equivalent to a load in the primary of:

(a) 5K

(b) 25K

(c) 200R

(d) 40R.

3. An 8 ohm loud speaker is correctly matched to the output impedance of a power amplifier using a 4:1 transformer. The amplifier output impedance is:

(a) 100 Ω

(b) 32 Ω

(c) 128 Ω

(d) 68 Ω.

4. A transformer rated at 250 V input on the primary provides an output of 50 V at 5 A on the secondary, what should be rating of a fuse link in the primary circuit?

(a) 1 A,

(b) 5 A,

(c) 0.2 A,

(d) 0.5 A.

5. A multi-secondary winding transformer is driven from a 250 V a.c. supply to provide 3 outputs of 5 V at 2 A, 10 V at 2 A and 5 V at 4 A, the rating of the primary winding would be:

(a) 250 V at 1 A,

(b) 250 V at 0.8 A,

(c) 250 V at 8 A,

(d) 250 V at 0.2 A.

6. Short circuited turns on the primary winding of a transformer cause:

(a) the transformer to behave as an open circuit,

(b) the primary current to fall to near zero,

(c) the secondary current to rise,

(d) the transformer to overheat.

Chapter 3

1. A transistor has g_m = 50 and h_{fe} = 500. Its base-emitter input resistance will be:

(a) 1K,

(b) 10K,

(e) 1M,

(d) 10M.

2. An N-channel MOSFET is to be used in a circuit in enhancement mode. In order to ensure that current flows in the channel, the FET:

(a) must have its gate connected to its source,

(b) must have its gate open-circuit,

(c) must have the gate at a negative bias relative to the source,

(d) must have the gate at a positive bias relative to the source.

3. A thyristor is used to switch a smoothed d.c. supply. Once the thyristor has been switched on, it can be turned off by:

(a) pulsing the gate negative,

(b) momentarily reducing the power supply to zero,

(c) pulsing the gate positive,

(d) applying AC to the gate.

4. Which of the following light-sensitive transducers needs no power supply?

(a) phototransistor,

(b) photoconductive cell,

(c) photodiode,

(d) photovoltaic cell.

5. A light meter uses a photodiode and an ordinary diode in the inputs to a balanced amplifier. The ordinary diode is used so as to:

(a) compensate for variations in temperature,

(b) compensate for variations in light,

(c) compensate for variations in power supply,

(d) compensate for variations in amplification.

6. A thermistor is used in a temperature controller. You would expect the circuit to make use of:

(a) a video amplifier,

(b) a balanced d.c. amplifier,

(c) an audio amplifier,

(d) an r.f. amplifier.

Chapter 4

1. A d.c. multimeter has a sensitivity of 20 000 ohms/volt, what is the total resistance presented when it is switched to the 100 volts range?

(a) 1.95M,

(b) 0.99M,

(c) 2M,

(d) 20M.

(2) The use of a digital voltmeter is often preferable to the use of an analogue voltmeter because:

(a) it has an infinite value of input resistance,

(b) it has a variable low input resistance,

(c) it has a negligibly small input resistance,

(d) it has a constant high input resistance.

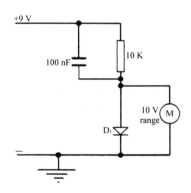

3. In the circuit, upper left, the meter reads 9 V on its 10 V range because of a fault in a component:

(a) the resistor is short circuit,

(b) the capacitor is short circuit,

(c) the diode is open circuit,

(d) the diode is short circuit.

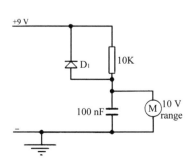

4. In the circuit, lower left, the meter reads zero volts on its 10 V range because of a component fault:

(a) the diode is open circuit,

(b) the capacitor is short circuit,

(c) the resistor is short circuit,

(d) the diode is short circuit.

5. A d.c. moving coil meter has a total resistance of 1000R and full scale deflection is produced by a current of 50 μA. What is the value of the multiplier (series resistance) needed to convert it to read 10 volts full scale deflection?

   (a) 199K,

   (b) 2001K,

   (c) 1.99M,

   (d) 99K.

6. A d.c. moving coil meter has a total resistance of 1000 Ω and a full scale deflection current of 50 μA. What is the value of the shunt (parallel resistance) needed to convert the meter to read 100 mA full scale deflection?

   (a) 0.707 Ω,

   (b) 1.414 Ω,

   (c) 0.637 Ω,

   (d) 0.5003 Ω.

Chapter 5

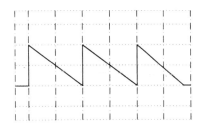

1. The graph, left, shows a waveform on the screen of a CRO with the timebase set to 1 ms/cm and the Y attenuator to 1 V/cm. The sawtooth wave is of:

   (a) 2 V p–p and 500 Hz frequency,

   (b) 2 V p–p and 2 kHz frequency,

   (c) 1 V p–p and 2 kHz frequency,

   (d) 1 V p–p and 500 Hz frequency.

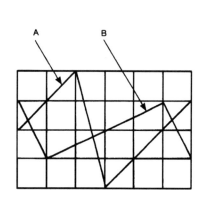

2. The graph, left, shows the waveforms on the screen of a CRO with the timebase set to 10 ms/cm and the Y attenuator to 5 V/cm. The frequency and peak-to-peak amplitude of waveform A is:

   (a) 50 Hz and 10 V,

   (b) 20 Hz and 20 V,

   (c) 20 Hz and 10 V,

   (d) 50 Hz and 20 V.

3. In the same graph with the same settings, waveform A leads waveform B by:

   (a) 20 ms,

   (b) 270°,

   (c) 90°,

   (d) 30 ms.

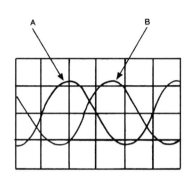

4. In the same graph with the same settings the frequency and periodic time of both waveforms is:

   (a) 50 ms and 20 Hz,

   (b) 20 ms and 50 Hz,

   (c) 20 ms and 20 Hz,

   (d) 50 ms and 50 Hz.

5. For the waveforms shown, left, if the X timebase is set to 100 μs/cm and the Y attenuator to 5 mV/cm, the frequency and peak-to-peak amplitudes of the waves are:

   (a) 400 Hz and 2.5 mV,

   (b) 2.5 kHz and 12.5 mV,

   (c) 400 Hz and 12.5 mV,

   (d) 2.5 kHz and 2.5 mV.

6. In the same graph as applied to question 5, waveform B leads waveform A by;

   (a) 100 μs,

   (b) 400 μs,

   (c) −135°

   (d) 135°.

Chapter 6

1. Comparing the operating principles of the BJT and FET devices:

   (a) the BJT is voltage driven and the FET current driven,

   (b) both are voltage driven,

   (c) the BJT is current driven and the FET voltage driven,

   (d) both are current driven.

2. A digital amplifier would operate in the:

   (a) cut-off region,

   (b) saturated region,

   (c) active region,

   (d) cut-off and saturated regions.

3. Which type of oscillator circuit would be most suitable for driving a low frequency signal generator?

   (a) Crystal controlled oscillator,

   (b) resistance/capacity controlled oscillator,

   (c) LC oscillator,

   (d) multivibrator oscillator.

4. When testing the condition of a capacitor, after checking its the value, the next most important property to resolve is its:

(a) loss resistance,

(b) Q factor,

(c) self inductance,

(d) frequency response.

5. An amplifier has an output of 6 dBW (6 dB relative to 1 watt), if the input signal level is 0.5 watts, the output level will be:

(a) 6 watts,

(b) 1.5 watts,

(c) 2 watts,

(d) 1 watt.

6. A test signal generator has an output at the 0 dB setting of 20 volts, what is the level of the output at the −20 dB setting?

(a) 2.5 volts,

(b) 1.5 volts,

(c) 0.5 volts,

(d) 2 volts.

Chapter 7

The first two questions apply to the screen shot, Figure 7.1, in Chapter 7.

1. The amplitude of the waveform (Figure 7.1) relative to a level of 1 volt (dBV) is approximately:

(a) 8 dB,

(b) −4 dB,

(c) −8 dB,

(d) 4 dB.

2. The value of the d.c. component in the waveform (Figure 7.1) is approximately;

(a) 8 volts,

(b) −0.4 volts,

(c) +0.4 volts,

(d) 0 volts.

3. If the maximum signal frequency to be examined is 100 MHz, the digital measuring system should be set to at least:

(a) 100 MS/s,

(b) 50 MS/s,

(c) 200 MS/s,

(d) 10 MS/s.

4. When a digital virtual instrument is set to a sample rate of 20 MS/s, the maximum signal frequency that can be processed is:

   (a) 10 MHz,

   (b) 20 MHz,

   (c) 40 MHz,

   (d) 30 MHz.

5. A 3-stage audio amplifier produces an output of 1 volt RMS for 10 mV RMS input. If the gains of the first and last stages are 1 8dB and 6 dB respectively, the gain of the intermediate stage is:

   (a) 18 dB,

   (b) 16 dB,

   (c) 14 dB,

   (d) 20 dB.

6. A human heart digital monitoring system is found to give a heart pulse rate of 72 per minute. What is this frequency in Hz and what should be the minimum sampling frequency setting?

   (a) 1.666 Hz and 0.833 S/s,

   (b) 0.883 Hz and 0.833 S/s,

   (c) 1.666 Hz and 1.666 S/s,

   (d) 0.833 Hz and 1.666 S/s.

Chapter 8

1. A bath tub diagram explains how the failure rate varies with:

   (a) working lifetime,

   (b) power rating,

   (c) power derating,

   (d) temperature derating.

2. Why are components not damaged when being passed through a wave solder bath?

   (a) Components are protected by the placement adhesive.

   (b) The PCB absorbs most of the excess temperature.

   (c) Immersion time is short.

   (d) The solder used has a very low melting point.

3. An electrolytic with a value of 8 µF and 100 V rating, is used with a 12 V power supply. The effect on the capacitor is to:

   (a) reduce its effective value,

   (b) increase its lifetime expectancy,

   (c) have no significant effect on its behaviour,

   (d) increase the risk of early failure.

4. The time duration of an after repair soak test can often be shortened by:
   (a) working at an increased but controlled temperature,
   (b) applying a higher than normal supply voltage,
   (c) using a lower than normal supply voltage,
   (d) providing the supply voltage via a Variac.

5. Desoldering, testing and refitting a multi-pinned IC can:
   (a) save expense,
   (b) lead to an early failure,
   (c) provide useful experience,
   (d) increase costs.

6. A manufacturer's quality assurance programme can:
   (a) increase the cost of a product,
   (b) provide employment for a technician,
   (c) increase the lifetime of a product,
   (d) reduce the cost of a product.

Chapter 9

1. A 10 metre length of flexible mains cable has a loop resistance of 0.01 Ω, what would be the expected loop resistance of 100 metres of similar cable?
   (a) 0.01 Ω,
   (b) 1.00 Ω,
   (c) 0.5 Ω,
   (d) 0.1 Ω.

2. The *Low Voltage* directive (LVD) applies to power supplies rated at:
   (a) from 75 to 1500 volts a.c.,
   (b) from 50 to 1000 volts a.c.,
   (c) less than 60 volts a.c.,
   (d) less than 60 volts d.c.

3. The insulation resistance between the casing and earth of a double insulated machine should be:
   (a) more than 7M,
   (b) more than 2M,
   (c) less than 7M,
   (d) less than 2M.

4. A 10 metre length of flexible mains cable has an insulation resistance between the live and earth conductors of 20M. What would be the expected value for a 100 metres length of such cable?

   (a) 0.2M,

   (b) 2M,

   (c) 20M,

   (d) 200M.

5. The earth bonding resistance on a Class 1 item of electrical equipment should be:

   (a) less than 7M,

   (b) less than 2M,

   (c) less than 0.5 Ω,

   (d) less than 0.1 Ω.

6. The COSHH regulations refer to:

   (a) the use of hazardous chemicals,

   (b) the use of lead free solder,

   (c) the training of first aiders,

   (d) the control of fire extinguisher services.

Chapter 10

1. A series regulator transistor:

   (a) has the same voltage drop across it as the load,

   (b) carries the same current as the load, but with a varying volts drop,

   (c) dissipates the same amount of power as the load,

   (d) has a constant voltage drop across it.

2. The reservoir capacitor of a power supply does **not**:

   (a) stabilize the output,

   (b) reduce the level of r.f. interference,

   (c) reduce the ripple,

   (d) increase the d.c. level.

3. A dummy load for testing a switched mode power supply can be:

   (a) an incandescent lamp bulb,

   (b) a high power low resistance load,

   (c) a high power high resistance load,

   (d) a transformer coupled load resistance of suitable value.

4. The main action of a power amplifier is to provide:

   (a) current gain,

(b) reduced distortion,

(c) voltage gain,

(d) reduced bandwidth.

5. A TV power supply uses thyristors in a bridge circuit for mains voltage at mains frequency. This type of circuit:

(a) allows much smaller smoothing capacitors to be used,

(b) always requires a transformer to be used,

(c) allows the voltage to be stabilized by controlling the thyristors,

(d) permits the chassis of the receiver to be isolated.

6. A typical switched mode power supply uses 40 kHz switching. This ensures that:

(a) the output will be well stabilized,

(b) smoothing will require only small capacitor values,

(c) output resistance will be low,

(d) no transformer will be required.

Chapter 11

1. Increasing the value of the load resistance on a single stage voltage amplifier will:

(a) increase the bias current,

(b) decrease the bias current,

(c) increase the gain,

(d) increase the bandwidth.

2. Feedback of signal from the collector of a transistor to the base of the same device will cause:

(a) reduced gain,

(b) more distortion,

(c) oscillations,

(d) reduced bandwidth.

3. A transformerless push–pull power amplifier must use:

(a) a stabilized power supply,

(b) a phase splitter,

(c) high resistance transistors,

(d) complementary transistors.

4. If a transistor amplifier is to be used at d.c. and very low frequencies, then it cannot make use of:

(a) negative feedback,

(b) differential amplification,

(c) capacitor coupling,

(d) balanced power supplies.

5. An ideal operational amplifier would have;
   (a) very high output impedance,
   (b) very high input impedance,
   (c) very low gain,
   (d) very low common mode rejection ratio.

6. At a virtual earth point there will be:
   (a) negligible d.c. voltage,
   (b) negligible feedback,
   (c) negligible offset voltage,
   (d) negligible a.c. voltage

Chapter 12

1. Which of the following is **not** a requirement for an oscillator circuit:
   (a) positive feedback,
   (b) negative feedback,
   (c) voltage gain,
   (d) current gain.

2. The Colpitts and Hartley circuits are examples of:
   (a) astable oscillators,
   (b) RC coupled oscillators,
   (c) sawtooth oscillators,
   (d) sine wave oscillators.

3. The important feature of a phase locked loop (PLL) is that:
   (a) it provides a stable output frequency,
   (b) its output frequency can be determined by a crystal,
   (c) the output frequency will be locked to an input frequency,
   (d) its output frequency is not affected by changes in supply voltage.

4. For a square wave input, a Miller integrator circuit will generate:
   (a) a sawtooth or triangular wave,
   (b) a sine wave,
   (c) a train of pulses,
   (d) another synchronized square wave.

5. An oscillator is an example of an amplifier that has:

(a) low gain and very wide bandwidth,

(b) high gain and very narrow bandwidth,

(c) high gain with very wide bandwidth,

(d) low gain with very narrow bandwidth.

6. An aperiodic oscillator employs:

(a) a crystal controlled circuit,

(b) an inductor/capacitor circuit,

(c) a surface acoustic wave (SAW) device circuit,

(d) a time constant circuit.

Chapter 13

1. The TTL logic level between +0.8 V and +1.4 V is known as the:

(a) threshold level,

(b) noise margin,

(c) nominal logic 0 level,

(d) nominal logic 1 level.

2. The Schottky diode threshold level is:

(a) 0.3 V,

(b) 0.6 V,

(c) 0.5 V,

(d) 0.1 V.

3. The rise and fall times of a waveform is defined as the time taken to change between the:

(a) 30 per cent and 70 per cent levels,

(b) 63 per cent and 37 per cent levels,

(c) 10 per cent and 90 per cent levels,

(d) −50 per cent and +50 per cent levels.

4. Tri-state logic devices operate with:

(a) 6 bit levels,

(b) 4 bit levels,

(c) 3 bit levels,

(d) 2 bit levels.

5. The propagation delay of a TTL logic chip is the time taken:

(a) for the output to respond to power up,

(b) for the output to change following an input signal,

(c) for the input to respond to power up,

(d) for the input to revert to its original level after the output changes state.

6. A logic circuit is described as combinational if:
   (a) its inputs are combined to produce the output,
   (b) its output is a set of combined signals,
   (c) the output is determined by the combination of inputs,
   (d) each input is a combination of signals.

Chapter 14

1. An analogue signal contains frequency components up to 2.5 MHz. What must be the minimum sampling rate for near error free transmissions?
   (a) 1.25 MS/s,
   (b) 5 MS/s,
   (c) 2.5 MS/s,
   (d) 7.5 MS/s.

2. A cable TV transmission system employs 256-QAM modulation. The number of bits per symbol is thus:
   (a) 8 bits,
   (b) 16 bits,
   (c) 32 bits,
   (d) 256 bits.

3. The packet structure of a digital transmission system allows:
   (a) data packets to be multiplexed,
   (b) for the addition of error control bits,
   (c) for an improved signal to noise ratio,
   (d) for a reduced bandwidth signal.

4. A reduction of bandwidth on a transmitted digital signal:
   (a) reduces the signal to noise ratio and increases the bit error rate,
   (b) increases the signal to noise ratio and increases the bit error rate,
   (c) increases the signal to noise ratio reduces the bit error rate,
   (d) reduces the signal to noise ratio reduces the bit error rate.

5. Why does the dispersion of square pulses occur in any transmission medium?
   (a) Because this is intentional in order to minimize the bandwidth of the signal.
   (b) Because this reduces the bandwidth requirements of the receiver.
   (c) Because all such systems have a low pass filter characteristic.
   (d) Because this is due to the addition of forward error control bits.

6. Why are digital signals encrypted before transmission?
   (a) To enable bit errors to be corrected.
   (b) To reduce the signal bandwidth.
   (c) To improve the signal to noise ratio.
   (d) To improve the security of the signals.

Chapter 15

1. A sequential circuit can best be described as:
   (a) a circuit containing flip-flops,
   (b) a circuit containing gates,
   (c) a circuit in which the inputs are in sequence,
   (d) a circuit in which the output is determined by the sequence of inputs.

2. The main disadvantage of the asynchronous or ripple type of counter is that:
   (a) it needs to be constructed from J-K flip-flops,
   (b) the last flip-flop cannot change state until all the preceding flip-flops have changed,
   (c) the pulses to be counted are applied to all the flip-flops,
   (d) it needs to be constructed from D-type flip-flops.

3. A shift register cannot be used for:
   (a) counting input pulses,
   (b) storing input pulses,
   (c) delaying input signals,
   (d) combining input signals.

4. A shift register designed to convert a data stream from serial to parallel format operates in the:
   (a) parallel in, serial out mode,
   (b) parallel in, parallel out mode,
   (c) serial in, parallel out mode,
   (d) serial in, serial out mode.

5. A 4-stage shift register counter with feedback to ensure that it resets at 1010 operates as a Modulo-N counter, where N is:
   (a) 16,
   (b) 10,
   (c) 8,
   (d) 2.

6. When a shift register is used as a memory device, it operates as:
   (a) a volatile device,
   (b) a multiplexer,
   (c) a non-volatile device,
   (d) a demultiplexer.

Chapter 16

1. After repairing any item of test equipment, the final act should be to:
   (a) recalibrate the instrument,
   (b) carry out tests to ensure that the instrument meets the requirements of LVD EN60950,
   (c) return it to service,
   (d) connect it to a power supply and check its functionality.

2. If a particular fault occurs frequently on a certain item of equipment, the service technician should:
   (a) keep quiet about as it could become a useful ego-boosting job,
   (b) note the solution in the service manual,
   (c) make out a report to advise the manufacturer of the problem,
   (d) record the occurrence in the accident book.

3. The plug on a certain 4-way cable has 2 pins that are longer than the others. This is:
   (a) to ensure that the power supply is connected before the signal lines,
   (b) to enable the cable to be connected to an IEEE488 bus connector,
   (c) because the opposite end has an RS232 connector for use with a parallel printer,
   (d) because these are connected to the mains plug interference filter.

4. Which of the following is not a common problem associated with the reliability of ICs:
   (a) pins short together during measurements,
   (b) incorrect voltage levels,
   (c) electrostatic discharges,
   (d) use of surface mount components.

5. What type of equipment should not be serviced in an ESD free environment whilst wearing an earthing strap:
   (a) mains connected equipment,
   (b) MOSFET populated circuit boards,
   (c) battery operated systems,
   (d) remote control units.

6. Which type of memory back-up battery should never be connected with reversed polarity?

(a) Nickel cadmium,

(b) Alkaline,

(c) Lithium ion,

(d) Manganese silver.

Answers to multiple-choice questions

Chapter 1	1(a); 2(c), 3(c), 4(d), 5(b) 6(c).
Chapter 2	1(c), 2(b), 3(b), 4(a), 5(d), 6(d).
Chapter 3	1(b), 2(c), 3(b), 4(d), 5(a), 6(b).
Chapter 4	1(c), 2(d), 3(c), 4(b), 5(a), 6(d).
Chapter 5	1(a), 2(b), 3(d), 4(a), 5(b), 6(c).
Chapter 6	1(c), 2(d), 3(b), 4(a), 5(c), 6(d).
Chapter 7	1(c), 2(d), 3(c), 4(a), 5(b), 6(d).
Chapter 8	1(a), 2(c), 3(d), 4(a), 5(b), 6(c).
Chapter 9	1(d), 2(b), 3(a), 4(b), 5(c), 6(a).
Chapter 10	1(b), 2(b), 3(a), 4(a), 5(c), 6(b).
Chapter 11	1(c), 2(a), 3(d), 4(c), 5(b), 6(d).
Chapter 12	1(b), 2(d), 3(c), 4(a), 5(b), 6(d).
Chapter 13	1(b), 2(a), 3(c), 4(d), 5(b), 6(c).
Chapter 14	1(b), 2(b), 3(a), 4(d), 5(c), 6(d).
Chapter 15	1(d), 2(b), 3(d), 4(c), 5(b), 6(a).
Chapter 16	1(b), 2(c), 3(a), 4(d), 5(a), 6(c).

Answers to Activity questions

Activity 1.1

16K, 3K4, 1K6, 318R, 159R, 80R, 53R. Non-linear and decreasing with frequency.

Activity 1.3

628R, 2K9, 6K3, 31K4 62K8, 126K, 188K. Graph is linear and increases with frequency.

INDEX